Praise for

The
DARKNESS
MANIFESTO

Named a Best Popular Science Book of the Year by
Waterstones and the *Financial Times*

"Urges us to reconsider our drive to dispel the dark."
—*The New York Times*

"Though the book is written as a sort of *Silent Spring* manifesto
against the ecological devastations of light pollution, its consid-
erable charm depends on the encyclopedic intensity with which
[Eklöf] evokes the hidden creatures of the night. . . . [The] book
is made most memorable by the sometimes wild eccentricities of
the life-forms it chronicles. Though his catalogue of catastrophe
is real, what one most remembers are the beasts in his bestiary."
—Adam Gopnik, *The New Yorker*

"Offers a powerful argument for turning down the lights and embracing the dark."

—*Undark Magazine*

"As a Swedish conservationist, Johan Eklöf urges us to think of light pollution as more than a nuisance that obscures our starry skies. In a series of well-researched vignettes, his message is a plea for nonhuman species: artificial lights disrupt migration patterns, mating rituals, pollination practices, insect biomass, and much more. Eklöf highlights the startling sprawl of these lesser-known consequences without evoking a hopeless or cynical tone. Instead, the book is a reflective reminder that our control of the world is as delicate as the smallest of species affected by it."

—*Scientific American*

"Poetic and philosophical . . . intimate and expansive."

—*Daily Telegraph* (UK)

"Though Mr. Eklöf loves both bats and darkness, his sensibility is more good-natured than gothic, rendering night not as an ordeal but an odyssey, rich in revelation and insight . . . lyrical . . . All of this might make for pretty grim reading, but Mr. Eklöf is really more of a rhapsodist than a scold, tempering peril with possibility."

—*The Wall Street Journal*

"A pleasure to read [and] a paean of praise for natural darkness."

—*Financial Times* (UK)

"What is lost when darkness disappears? Stargazing, flowers that unfold by moonlight, phosphorescence in the sea, and something deeper yet. With extraordinary insight, Johan Eklöf explores the influence of night on nature, on cities, and in our connections to one another. A timely synthesis of the latest research on light pollution becomes an argument for the preservation of shadows. *The Darkness Manifesto* is a scintillating read by a conservationist of true literary flair, who has spent long hours tuning his attention to twilight and nocturnal life."

—Rebecca Giggs,
award-winning author of *Fathoms*

"An urgent and erudite hymn to the night, composed by a scientist with the soul of a poet."

—Chloe Aridjis, author of *Book of Clouds*

"Eklof's excellent book covers a lot of ground. It shaped and sharpened my thinking on a subject I knew I was supposed to feel something about. In the week after I read it, I found myself saying, again and again, at the dinner table (six bulbs above it where one would suffice), 'Yes, children, but did you know . . . ?' I can think of no better praise than that."

—Christopher Kemp, *Science*

"[An] eye-opening treatise on light pollution . . . Urgent and vivid, this account shines."

—*Publishers Weekly*

"Absolutely wonderful, full of graceful insight and gentle persuasion."

—Chris Goodall,
author of *What We Need to Do Now*

"In this superb book by a world-leading ecosystem ecologist, Eklöf takes us on an exciting journey spanning multiple fascinating areas of nocturnal biology. . . . This is a critically important must-read for all who have an interest in the health of our planet. Which should be all of us."

—Russell Foster,
director of the Sleep and Circadian
Neuroscience Institute and head of the
Nuffield Laboratory of Ophthalmology
at the University of Oxford

The

DARKNESS
MANIFESTO

*On Light Pollution, Night Ecology, and
the Ancient Rhythms That Sustain Life*

JOHAN EKLÖF

Translated from the Swedish
by Elizabeth DeNoma

SCRIBNER

New York   London   Toronto   Sydney   New Delhi

Scribner
An Imprint of Simon & Schuster, LLC
1230 Avenue of the Americas
New York, NY 10020

First Scribner trade paperback edition February 2024

SCRIBNER and design are registered trademarks of The Gale Group, Inc.,
used under license by Simon & Schuster, LLC, the publisher of this work.

Simon & Schuster: Celebrating 100 Years of Publishing in 2024

For information about special discounts for bulk purchases,
please contact Simon & Schuster Special Sales at 1-866-506-1949
or business@simonandschuster.com.

The Simon & Schuster Speakers Bureau can bring authors to
your live event. For more information or to book an event,
contact the Simon & Schuster Speakers Bureau at 1-866-248-3049
or visit our website at www.simonspeakers.com.

Interior design by Kyle Kabel

Manufactured in the United States of America

1   3   5   7   9   10   8   6   4   2

Library of Congress Cataloging-in-Publication Data
Names: Eklöf, Johan, author. | DeNoma, Elizabeth, translator.
Title: The darkness manifesto : on light pollution, night ecology, and the ancient rhythms
that sustain life / Johan Eklöf, translated from the Swedish by Elizabeth DeNoma.
Description: New York : Scribner, 2023. | Tranlsation of: Mörkermanifestet.
Sweden : Natur & Kultur, c2020. | Includes bibliographical references and index.
Identifiers: LCCN 2022037888 (print) | LCCN 2022037889 (ebook) |
ISBN 9781668000892 | ISBN 9781668000915 (ebook)
Subjects: LCSH: Night—Physiological effect. | Light pollution. | Light and darkness—
Physiological effect. Classification: LCC TD195.L52 E4413 2023 (print) |
LCC TD195.L52 (ebook) | DDC 363.7—dc23/eng/20221101
LC record available at https://lccn.loc.gov/2022037888
LC ebook record available at https://lccn.loc.gov/2022037889

ISBN 978-1-6680-0089-2
ISBN 978-1-6680-0090-8 (pbk)
ISBN 978-1-6680-0091-5 (ebook)

Contents

CONTENTS

PART III **Humanity and the Cosmic Light**

CONTENTS

Introduction:
The Disappearing Night

My flashlight sweeps over a demon painted black with bat wings and a snake for a tail. The creature looks as if it's being thrown backward, light radiating from its mouth, as if it has tried to swallow the light, but can no longer resist its power. The creature of darkness is dying. I am in an eighteenth-century church in Sweden, painted with biblical themes, and in the far back you can find the most horrific devils and demons, put there to remind us of the torments of hell. But perhaps the church painter also wanted to tell us that we can overcome the dangers of darkness. From the church's perspective, bats are the devil's minions, filthy animals that are symbols of both literal and philosophical darkness, in opposition to the light of God. So, it's a bit ironic that churches have so often become the nesting places of these creatures.

I continue exploring the church, climbing up a flight of stairs and stepping through a little door into the attic. On the old wooden floorboards are piles of guano and severed butterfly wings, a clear sign that the church is inhabited by the brown long-eared bat. The dusk that flickers in through the slats grows weaker, and outside the sky turns navy blue. The humid night air entering the attic carries a pleasant smell of freshly cut grass, tar, and sun-warmed wood. The bats are unwilling to show themselves from under the eaves this early in the evening, so I go outside to meet them in the cemetery and watch them alight into the summer night.

One after another they take off headfirst from the roof to the nearest tree and its protective dark shadows. In a fitful dance, they glide, inaudible to the human ear, by the red-painted wooden church, alongside the hedges and around the treetops, searching for insects. Soon the bats will be gone, swallowed by the night.

Swedish churches and their outbuildings have frequently been tended in the same way for centuries and have grown to be important havens for animals and plants in an ever-changing world. Year after year, long-eared bats have moved into turrets and attics in the early summer to give birth to new generations. In the 1980s, two-thirds of the churches in southwest Sweden had their own bat colony. Today, forty years later, research I've done with my colleagues shows that this number has been reduced by a third due to light pollution and other factors. Because the churches all glow like carnivals in the night. District after district has installed modern floodlights to show the architecture it's proud of, all the while the animals—who have for centuries found safety in the darkness of the church towers and who have for 70 million years made the night their abode—are slowly but surely vanishing from these places, maybe completely.

Sitting in the cemetery in the July night, I'm not only in the company of bats. I can see a hedgehog, beetles making their way up through the grass toward the starry sky, and above the headstones caddis flies dancing like spirits. I start to relax in the gathering darkness, as all the impressions of the day are traded for more subtle experiences, and my eyes become slowly acclimated to the night. I've entered another dimension that few others ever take the time to visit.

It is not just the bats and I who enjoy the darkness. Most mammals are more active at twilight, such as the hedgehog that keeps me company in this late hour. Half of the insects on this planet are nocturnal, and for the last couple of years we have been drowning in alarming reports regarding their disappearance. Forestry, environmental toxins, large-scale farming, and climate change—many causes are mentioned but little is said about light, even though the light-sensitive moth is one of those most disappearing. Moths looking for nectar in the darkness are easily confused by all the lights. They either don't fly at all, believing that dawn is about to break, or they're disoriented by the beams of light when they try to navigate by using the moon. Exhausted, they die or get killed by predators, without having fulfilled their nocturnal mission, and thus fewer plants are pollinated. Many of us have probably seen the phenomenon out on our porch or under a streetlight—the brighter the lights, the greater the attraction. The light lures insects from forests and villages, from the countryside and into cities, depleting entire ecosystems.

Mossebo Church may lack floodlighting, but some light still reaches this place. A few lights are alongside the walking paths, and in the sky a faint orange glow can be seen coming from the nearest villages. This is *light pollution*—a collective term used for light that is regarded as superfluous, but which still has a great impact on our lives and our ecosystems.

The term was coined by astronomers, but is today used by ecologists, physiologists, and neurologists who study the effects of the disappearing night. It is no longer just a question of stars and insects. It is about all living things, including we humans. Ever since the birth of our planet, day has been followed by night, and every cell in every living organism has built-in machinery working in harmony with that rhythm. The natural light calibrates our inner circadian rhythm and controls hormones and bodily processes.

Up until about 150 years ago, when the light bulb was invented, these processes were allowed to develop slowly and without disturbance. But today, streetlights and floodlights ominously supersede natural night light and disturb this ancient circadian rhythm. The artificial light, the polluted light, is now dominant—light that causes birds to sing in the middle of the night, sends turtle babies in the wrong direction, and prevents the mating rituals of coral in reefs, which take place under the light of the moon.

Humanity's desire to illuminate the world makes Earth, viewed from space, seem to glow in the night. Every city and every street is visible a long way out in the cosmic darkness, which is perhaps one of the most obvious signs that we have entered a new era: the Anthropocene, the time of humans. Beneath the illuminated sky in the lit-up cities we have created, we can no longer see any stars, and many of us don't remember what the Milky Way looks like. We are missing out on one of nature's grand treasures: the spectacles of the sky with its breathtaking perspectives, its falling stars and, on occasion, its strikingly beautiful northern lights.

Light pollution is still a term unknown to many, but it's an exploding field of research, and light will probably soon be as strictly regulated as noise. The LED light, the modern diode, which has enabled the explosion of lighting in private gardens and industrial parking lots, could also be a solution to the problem. Light and darkness are not a matter of black or white. We can program and dim the artificial light and adapt it to more natural conditions. If we want to.

With this book, I examine the impact that darkness and the night have on all living creatures. In a number of concise chapters, I'll share my experiences and thoughts stemming from my twenty years in the service of the night, as a bat researcher, traveler, and friend of the darkness. I hope that this book will inspire others, function as a reminder of the importance of letting the night be a part of our lives, and give insight into how much damage artificial light can do—be a challenge and a manifesto for the natural darkness.

PART I

Light Pollution

The Cycle of Darkness

Mimosa pudica has an odd quality. The plant is sensitive to being touched, and if you brush against its leaves, they close up like an umbrella and seem to whither before your eyes. The same thing happens at night. Every morning the leaves open and turn like satellite dishes to capture the sunlight. The French scientist Jean-Jacques d'Ortous de Mairan (1678–1771) placed a plant in constant darkness and discovered that the leaves still opened when it was day outside, even though the plant saw nothing of the sun. He interpreted this to mean that the plant still felt the presence of the sun. How this could be de Mairan never managed to explain.

Only during the second half of the last century, with breakthroughs in genetics, was the mystery solved. In the 1960s, the biologist and geneticist Michael W. Young had begun to ponder the mimosa plant's and other plants' behavior during different times of the day, and from this pondering a lifelong interest in the biological clock was born. In 2017, Young, along with

Jeffrey C. Hall and Michael Rosbash, won the Nobel Prize in Physiology or Medicine. They had succeeded in isolating the gene that controls the rhythm in all living things, from bacterium to human being. The *circadian* rhythm, which can be thought of as our internal food and sleep clock, has been with us since the beginning of time, following the day's natural progression from darkness to light and back again.

Through billions of years—Earth is 4.5 billion years old—our planet has changed form, slowly or in sudden events. Mountain ranges and seas are built, rivers are moved, and species are born and die out. Not even the magnetic poles constitute fixed points. Right now, the magnetic north pole is moving eastward, from northern Canada toward Siberia, at a speed of seven miles a year. But one thing has remained more or less constant: the alternation between day and night, between light and darkness. The sun has always gone down in the west and risen again in the east, and in between those moments it has been night.

The length of a day has not always been the same. Modern atomic clocks tell us that the earth's rotation is slowly decreasing, and the days are becoming longer. A little bit longer period of daylight, a little bit longer night. The day's rate of change is not dramatic, amounting to barely two milliseconds per century. But if the length of the day has always changed at the same rate, the earth's first life-forms, living more than 3 billion years ago, experienced a day that was only half as long as ours.

There are many theories about where this first life, a life that was not much more than self-copying molecules, arose: in the deep sea, under thick ice, far inside mountain crevices in

THE CYCLE OF DARKNESS

a mass of mud, or potentially even in some other place in the universe. But wherever life first came about, the first single-cell organisms developed quickly and found new possibilities in the unexplored world.

And soon cyanobacteria, organisms with the ability to make use of sunlight to create oxygen, spread over the world's oceans. Every morning when the sun's first rays warmed the surface of the water, cyanobacteria, which we also know as blue-green algae, could gather the light's energy and fill the atmosphere with oxygen. They played a crucial role for the atmosphere's chemical composition, so that animal life, including humans, could develop. Cyanobacteria's inner machinery laid the foundation for plant development and photosynthesis, and their rhythm has propagated itself through generation after generation.

The earth's first multicellular life saw the light of day 620 million years ago, when a day was around twenty-two hours long. Or rather, they didn't literally see the light of day. It would still take millions of years before eyes or any other truly advanced senses existed. During this time, life-forms unique to their time thrived, organisms that died out more than half a billion years ago. But for millions of years, they could live a quiet life on luxuriant carpets of algae, without risk of predators and without needing to move a millimeter. Every day sunshine trickled through the surface water and changed character on its way to the deep. Every evening effects from the light stopped, and the natural night took hold. Life adapted to these shifts.

The biological clock, our circadian rhythm, is ancient, shared, and completely fundamental. Everything living today has developed in a world where the conditions are changed over the day and over the year. Our bodies simply expect light and dark in recurrent cycles of longer or shorter nights. Every organism makes use of the preprogrammed clock in different ways, as when the mimosa plant collapses its leaves at night, the butterfly orchid wakes up to life instead and boosts its scent to attract moths. The bee and other daytime insects end their shift, and the night pollinators begin. All employ the same foundational mechanism, regardless of species, habitat, or life cycle, from the 2.5-billion-year-old cyanobacteria to bats to humans.

It's light and darkness that calibrate the biological clock. Without information about changes in the surroundings, the inner mechanism continues to pulse in regular rhythm for about a day. The morning light signals that the cycle must begin again from zero, that a new day has just begun. The clock continues over the day, through dusk toward night, the whole time with input from the sun's varying light. The artificial light from lamps, headlights, and floodlights is not in this equation and risks, to put it mildly, creating disorder in the system.

Experiences in Darkness

I usually begin my nightly inventory by sitting down in a peaceful place, preferably near some water. I pour a cup of coffee from my thermos and let my mind passively take in impressions of twilight. The steam from the thermos mingles with the fog over the water as darkness falls and still air cools near the surface. The birdsong grows more sparse, the long-horned grasshopper's hoarse sound becomes sharper, and a dark green backdrop builds in the forest. During summer in Scandinavia, the day's transformation to night can carry on a long while, as a subtle displacement of light and activity, where the day's animals meet the night's, where the songbirds' warble hardly falls off before the woodcock in swift flight demonstrates that twilight has arrived. In the tropics the change occurs rapidly, as if scenes on a theater stage have been moved. The spotlight is replaced with shadow, and the stage and audience remain the same, but the actors are new.

Sometimes it takes a while for the bats to show up. The infinite is primary here. I want to believe that I work more effectively in the long run if I let natural pauses happen, if I let the darkness fall in its own time. The natural experience does not necessarily make me a better field-worker, but one who is more harmonious. Had I allowed internet surfing and my cell phone to disturb me with their light and audible pings, I would have lost both focus and my night vision.

I am unwilling to lose my night vision and seldom even use a headlamp, at least not outside. Otherwise I would not have seen the ground beetles hunting small insects or the spiderwebs glittering in their special way in the moonlight. A lot would have passed me by, such as the slugs moving themselves along and the mushrooms lighting up. Indeed, some fungi have biolumi-nescence, the same quality that gets sea fire and glowworms to gleam in the dark. With this light they attract flies, beetles, and ants, which spread the fungi's spores. The phenomenon is most common in the tropics, but here in Sweden we also have an illu-minating fungus, honey fungus, whose mycelia—the fungus's threadlike network—glow a dim green. People in earlier eras were said to use wood from oaks overtaken by fungal mycelia to show them the way at night. Maybe mushrooms stand out like bright lanterns for animals with better night vision than we humans have.

It is fascinating to imagine how nocturnal animals experience their existence in the dark, how their brains interpret sensory stimuli. In my vicinity hundreds of normally invisible white flowers called Nottingham catchfly glitter when the moon shows itself. It is subtly beautiful, but for animals with sensitivity to the ultraviolet spectrum the ground shines like a fluorescent dance floor. As humans—with our senses' limitations—though we know about these animals' visual faculties, we can never understand the real experience. Filters in cameras or visual enhancement through other machines let us have an inkling, but we can never completely see with the eyes of insects or cats. The philosopher Thomas Nagel wrote in the 1970s in his famous essay "What Is It Like to Be a Bat?" that human language can describe what it's like to be a bat as little as it can describe what it's like to be an extraterrestrial. Only individuals from the same species can understand one another's experiences, and if we extend Nagel's reasoning, we cannot know what it like to be another human being either. We only have our own senses, filters, and interpretations.

But if you remove yourself from the commuter lane, sit down as an observer, and let the darkness meet you, proximity to nocturnal life nevertheless makes it more striking. Senses other than sight take hold, and slowly the sounds and the smells change, the air becomes damp against your skin. A whip-poor-will, a bird of twilight, flies past with a suggestive and unmistakable drone. Some frogs croak, a far-off black-throated loon calls out its melancholy verse, and at a distance a splash in still water can be heard. Gradually night vision also increases, and you get an idea how flowers of the dark come to life, such as white campion, butterfly orchid, and night-flowering catchfly. They release their scent molecules and send spores with the wind for nocturnal pollinators to follow. During early summer's drawn-out twilight the lilac comes most fully into its own, and it's said that a person born around midnight can see the ghosts in the lilac's silhouette

9

on Sundays. In August, the smell of wild capriole takes over the summer night, and owlet moths are drawn toward the plant's blossoms thanks to the scent trail. With their long, sucking proboscis, the moths slake their thirst on nectar and pollinate the plant. Moths have the animal world's most exceptional sense of smell and can capture separate scent molecules with their antennae and find a flower or partner from several miles away. Allow yourself to sit outside during twilight and you'll soon get a feeling for the invisible scent trails by observing the moths' strenuous flights. Moths have shown themselves to be at least as important as pollinators as the diurnal bees, and moths even visit more different kinds of flowers than do bees, something of invaluable importance for keeping our ecosystem intact and thriving.

One of the moths I'm observing suddenly takes a deep dive toward the ground, then pulls an acrobatic loop to return to its scent trail again. Moths have developed hearing to listen for the sounds from the very bats I am here to inventory. So, the moths' sudden jerks are flights from the enemy. My ultrasound detector makes the bats' sounds audible for human beings, and these rattle like exploding popcorn. The closer the moth, the faster the bats cry out their search pulses to locate the prey. Moths veer and feint in a duel under the night sky, accompanied by a measured rhythm. On the ground, several beetles rush forward. Leaves rustle and soon two cockchafers rise in a mating dance. For a moment, the buzz of their beating wings overpowers the sounds from the ultrasound detector.

No less than a third of all vertebrates and almost two-thirds of all invertebrates are nocturnal, and so most of nature's activity—mating, hunting, decomposing, and pollinating—occurs after we humans fall asleep at night. As a bat researcher, I'm regularly reminded how little we still know about the night and its secrets. About the bats coordinating flight around trees, how in a microsecond they determine how the landscape around them looks only by using sound and its echo. The darkness is not the world of humans. We're only visitors.

Illuminated Planet

The bat, the whip-poor-will, and the cockchafer all belong to twilight, while the human being, to the utmost degree, is diurnal. We humans are in many ways completely dependent on visual sensory impressions, so therefore light means safety for us. So it isn't strange that we tend to want to light up our existence, and with electricity's and the light bulb's triumphant march across the world in the last 150 years, and now with the revolutionary diode lamps, this illumination is occurring at an even higher rate. We light the yards around homes, streets, industrial areas, and parking lots with lamps, floodlights, and strings of lights, often with safety in mind. In the school parking lot, a few hundred yards from where I live, they've put up about fifty lampposts. That's about one lamppost per twelve square yards of asphalt, mostly for the enjoyment of young people, who drive there to have a place to hang out at night. And it looks the same every-where else: light shines from empty offices, in vacant parking lots, and on the facades of warehouses along our motorways.

Human beings have extended their day while forcing out the night's inhabitants.

If you view satellite pictures of Earth at night today, they show a planet that glows. All the world's densely populated areas form brightly lit spots that can be seen far out into space. Lighted roadways bind cities together in a shining network, and the most densely populated parts make a single haze of light. Satellite pictures show concretely how the urbanized world spreads out, and this spread is perhaps one of the strongest symbols of what is called the Anthropocene. This concept was coined in the 1980s and was suggested later by the Dutch chemist and Nobel Prize winner Paul Crutzen to designate the age in which we live. To name a new geological epoch after human influence over the world isn't a new thought. The idea can be traced back to the 1860s and George Perkins Marsh (1801–82), an American politician, diplomat, and linguist who, somewhat unexpectedly, came to be a foundational figure in an early environmental movement. Inspired by his 1864 book, *Man and Nature; or, Physical Geography as Modified by Human Action*, people over the next two decades made a succession of attempts to name the ruling epoch after humans, given their virulent effects on the environment. But only now have the ideas about the Anthropocene really taken hold.

Night satellite pictures give a clear view of how modern human activity is spreading in time and space. Despite the much good that technological developments have done for humans—with their concrete as well as symbolic illuminations— in their tracks plainly follow energy wastefulness, rampant consumerism, and ecological degradation. What we call light pollution—unnecessary artificial light—changes nature and has hitherto been an underappreciated example of the Anthropocene. While artificial lighting today makes up just a tenth of our combined energy usage, only an extremely small part of that light is of actual benefit to us. Most of it spills out into the sky instead of lighting walkways and outer doors as intended. Research in Europe and the United States shows that badly directed and unnecessarily strong lights cause pollution that is equivalent to the carbon dioxide emissions of nearly 20 million cars. In 2017, light pollution was minimally estimated to increase by 2 percent globally each year.

One of the reasons for our eagerness to illuminate our planet so persistently is without a doubt our nyctophobia, fear of the dark. To be afraid of the dark lies in our genetic, as well as our cultural, heritage. It is altogether natural, and exactly like many other fears and reactions, it has a survival value. Our sense of sight adapts so that we see decently in the dark, but slowly. It takes at least half an hour for the right pigment to build up in our eyes when the daylight's bombardment of photons has begun to decrease, and a little while more before we reach our maximum light sensitivity, before we can orient ourselves in the dark. And the heightened sensitivity to darkness can be undone in an instant. One look at a streetlight, a cell phone powering on, or a passing car with headlights breaks down the rhodopsin—our light-sensitive visual pigment—which falls down like a house of cards, and the eye is forced to begin again.

In our cities today, it is almost impossible to establish true night vision, for there are far too many points of light, which effectively hinder the buildup of rhodopsin. In Hong Kong and Singapore, which are considered the earth's brightest cities, or maybe most specifically the most light polluted, there's barely a street corner where it's dark enough to call forth the eye's natural night vision. People in Hong Kong sleep under a night sky that is twelve hundred times brighter than an unilluminated sky, and if you were raised in Singapore, you've likely never experienced night vision. This applies to more and more of us who live in cities, wherever in the world we find ourselves.

The loss of the experience of night can possibly seem nostalgic and tangential. But there's a lot of research showing that the human being in the Anthropocene experiences strongly negative effects from too much artificial light. The light disturbs our biological clock, causing sleeping difficulties, depression, and obesity. Some published studies show that certain forms of cancer can be a direct result of too much light at night, but we'll talk more about all of this later.

The Vacuum Cleaner Effect

A moth steers toward a shimmering waterfall and disappears into the body of water. Soon another comes along, and before long a whole line of them. None of them hesitates or stops: they continue straight into the rushing water.

The observation was made at a waterfall in Iceland's Skjál-fandafljót River in the 1800s. What enticed the moths that night wasn't the need to cool off, nor was it their inevitable death in the water. It was the shine and glitter from the fall's water droplets, their hypnotic power of attraction, that drew the moths in. The observer—George John Romanes (1848–94), philosopher, psychologist, and biologist—studied instincts in both humans and animals and was fascinated by how the smallest light—the weak shine from a matchstick or a glittering droplet of water—could get insects off course. Romanes was working at Oxford University and a close friend of Charles Darwin's. Romanes was also an eager spokesman for Darwin's theories and was predicted to be his successor in defending evolution's throne.

Unfortunately, George John Romanes died at the early age of forty-six, gradually landing in the shadow of other biologists working in the field in the next century. But his ideas about animal instinct in his works *Mental Evolution in Animals* and *Mental Evolution in Man* have had a large influence within both zoology and psychology. Just as how Romanes made note of moths being drawn to the shimmering body of water from the Skjálfandafljót, most of us have sometime watched insects drawn to a light and observed how they come closer, little by little in ever tighter circles, to finally fall straight down in the center of the light source.

In 2001, I took part in a workshop about bats in Krau Wildlife Reserve in Malaysia's interior forest. As a young graduate student, halfway through my dissertation, I didn't want to miss the experience. A local TV crew was on location, following one of the domestic researchers in his work with bats. One evening, during dinner, one of the film crew's large lights was left on, directed up toward the sky. This created a compact column of light in the dark, humid rain forest air and showed with tremendous clarity what happens to insects around a light source. A heavy stream of moths, caddis flies, mosquitoes, beetles, crickets, and all manner of more or less obscure insects were caught in the column of light, and one by one the bugs danced in a spiral down toward the light. Well, not all of them. One opportunistic praying mantis, having landed on the edge of the light, was now raking in prey after prey. The praying mantis had transformed the film crew's light to its own private trap, and I sat a long time studying its seemingly conscious effort.

On the south end of the Strip, Las Vegas's most famous boulevard, a light installation towers overhead that functions in much the same way. Atop the Luxor Hotel and Casino sits what is America's and likely the world's most powerful stream of illumination—Luxor Sky Beam. With the help of a complex of curved mirrors and thirty-nine xenon lights at seven thousand watts each, a beam points right into space and can be seen nearly forty-five miles away, at least if you are at the cruising altitude of an airplane. The strength of the light is equivalent to the light of 42 billion candles. If you're going to stand out in one of the world's brightest cities, it doesn't do to skimp on power.

After an unusually damp 2019—at least by Nevada's standards—an enormous grasshopper migration was triggered in the area. Grasshopper swarms of this kind are nothing unusual, and the same thing happened in East Africa a half year later. Many species of grasshoppers happily migrate in large numbers, especially after seasons with a lot of rain. The grasshoppers can quickly create enormous populations, and when these reach a certain size, the grasshoppers' hormone systems tell them it is time to move. Their massive numbers not only form a spectacle, but can also present great societal problems, not least by destroying crops. It's difficult not to draw connections to the Bible in this case: a hotel in the form of an Egyptian pyramid, a sin city characterized by gaming and gambling, and a swarm of millions of grasshoppers that sweep in from the surrounding desert, as in one of God's ten plagues in the book of Exodus. Social media were full of fantastic film clips and commentary in July of 2019, when the invasion reached its climax.

The grasshoppers normally migrate at night, and every evening Nevada's meteorologists could see on their radar screens the swarms approach Las Vegas. All of the city's lights, including advertising screens and neon signs, were like magnets for the grasshoppers, and worst of all was the Luxor Sky Beam, which attracted insects all the way from Arizona. In insect circles people talk about a "vacuum cleaner effect." And precisely like the Luxor Sky Beam and the light I studied in the Malaysian forest, every streetlight, every porch light, and every illuminated facade is a seductive magnet for insects. On a larger scale, cities attract insects from rural areas, which leads to changes in the entire ecosystem.

The vacuum cleaner effect has long been used by entomologists to capture insects with light traps. These consist of a light and a box, and when the insects come flying, they are captured in a funnel and sealed in. This kind of trap sat on the roof of the Swedish Museum of Natural History between 1990 and 2007. Every year the lamp was visited by over 200 kinds of butterflies, and altogether during the seventeen years no fewer than 740 different species, including beetles and heteropterans. Over the years, the composition of species did not much change. But if those researchers had notated the number of individuals of each species and weighed them—measured their biomass— they could probably have perceived a trend. In Germany, the same type of measurements were introduced a year earlier in a study and were of a considerably larger scope. In more than sixty different nature reserves, insects were captured, identified by species, and weighed. And in 2013 the first warning came.

But not until four years later, after further data analysis, did the news reach the rest of the world. It then spread quickly via social media with selling subject lines such as "Harmageddon" and "Insect Collapse"; the biomass of the insects had decreased by 75 percent! The results were published in Open Access, available for all who wanted to draw their own conclusions, reinterpret the statistics, or review competing research. But the conclusion is clear: the numbers of insects are decreasing. The reasons for insect death are many, from urbanization and global warming to the use of insecticides, large-scale farming, single-crop cultivation, and disappearing forests. Probably all these factors play a role. But for anyone who's ever seen an insect react to light, it is obvious that light pollution is a major cause.

Extinguished Mating Impulse

No one knows exactly how many insect species there are in the world; it's in the millions, and new ones are constantly being discovered. In Sweden and Norway alone, sixteen hundred completely new species of insects have been identified in the last decade. In the tropics, every insect inventory leads to new discoveries, and many species presumably die off before we even have time to discover them.

Half of all species of insects are nocturnal and need at least several hours of continuous dark to obtain food and find a mate. The night's limited light protects these insects, and the pale glow from stars and the moon is central for their navigation and hormonal systems. Disturbances in the natural oscillation between light and dark is therefore a threat to the night insects' very existence.

To navigate the night, the majority of insects use the stars, the moon, or so-called polarized light. A moth during flight in darkness keeps a straight course by maintaining contact with moonlight, the brightest source of natural light in the night sky. When the moth instead comes upon a man-made light, an unnatural element that is infinitely closer than the moon, the moth slowly turns toward the light to continue on the same bearing. This results in a spiraling flight closer and closer to the light.

When insects have been caught in a hypnotizing light, they stay there. Many of them die before dawn, sometimes of sheer exhaustion. When the light eventually goes out, often at the same time the sun returns, the surviving insects have hardly moved at all and haven't achieved their night's goals. They haven't gotten their nectar (and transported the plants' pollen around), haven't found a partner, and haven't laid any eggs.

Insects frequently employ polarized light to find their way through the night's darkness, but maybe even humans have made use of this same trick. In the Icelandic sagas, for example, the Vikings navigated the seas by using a crystal, a sunstone. The stone revealed an unseen pattern in the sky, formed by the light from the sun's rays. No matter what the weather, the Vikings could see where the sun was as long as they looked through the crystal. Archaeologists have never found such a sunstone, but in theory it should function.

Light waves undulate in all directions, not just up and down or side to side. The directions are evenly divided as long as the light travels unhindered. But when the photons meet the molecules and particles in the air, or when the light passes through, for example, a water surface, some of the directions are filtered out, the light becomes *polarized*, and then the light waves swing, or vibrate, more in one direction than in others. This is happening all the time, and across the sky a pattern of light is formed that swings in different directions and has been polarized to different degrees. As the sun goes down, the polarization changes, and at dawn and dusk the pattern is least complex. It's as if the sun pulls its rays with it over the edge of the earth and leaves trails on the evening sky, trails that are both a compass and a clock. We human beings don't see this with our naked eye, but insects do, and furthermore, they use it to orient themselves.

We've long known that bees benefit from polarized light, but more recently we have discovered that a long line of insects, spiders, crustaceans, and even birds make use of the optical compass. Also, when the sun has long since sunk below the horizon, the moon can give the same effect, although its glow is forty thousand times weaker than the sun's.

The dung beetle is one of the best at using the moon's barely discernible light patterns in the sky. Dung beetles come in many different varieties, with some sixty species in Sweden alone. Most known are probably the dung beetles that make balls of animal droppings, which they then roll away with them over the African savanna. Using its back legs, the beetle assiduously pushes the ball to its nest. The ball can weigh many times the beetle's own weight, and a ball full of nourishment is a booty for other dung beetles, which is reason to rush it home. To find the closest and fastest way, dung beetles navigate with help from the moon's polarized light in the night sky, and even the weakest light from the sliver of a new moon can lead them in the right direction. They have such great sensitivity for nuanced differences in this light that even in environments close to big cities where weak traces of light spill out from streets and houses, the beetles can find their way. But in that case the moon must be full, for otherwise the trails in the sky are hidden, even for dung beetles. For safety's sake, they also make use of the stars to orient themselves in open landscapes. By climbing up on their ball, turning toward space, and performing a little dance, they create for themselves a snapshot of the night sky, like an astrophoto of the evening's heavenly pattern. Giant crab spiders in desert areas do this in a similar fashion. By calmly watching the night sky with their eight eyes, they create a picture of the horizon and the position of the stars, a star map, which helps them find their way among the desolate sand dunes.

When the dung beetle has sculpted its ball of manure, memorized the night's star images, and determined its direction, it rolls its ball home to safety in a straight line. In ancient Egypt, people believed that the dung beetles, the scarabs, laid their eggs in the ball, and therefore in that culture they became holy symbols for fertility. The dung beetles' journey was likened to the sun's path across the heavens, which is why Khepri, the morning guise of the sun god, Ra, is usually drawn with a dung-beetle head.

When light strikes a water surface, the light waves are changed and reflected in a distinct polarized pattern, by which caddis flies, diving beetles, and other water insects can navigate to find bodies of water. But artificial light can create false water surfaces. Asphalt, concrete, glass, and the glossy coatings on cars—all reflect light in a water-like way, and the artificial lights from houses, shopping centers, and industries strengthen the effect. I have found diving beetles on the hood of my car and seen mayflies land in parking lots to lay eggs. The mayflies' short adult life largely has no other purpose than to give the next generation good conditions in which to thrive, by laying eggs on water. If we instead distract the insects to industrial areas or large parking lots, entire populations can be knocked out overnight.

Aside from disturbing the insects' natural navigation and, in the worst cases, luring them to complete inactivity or death, artificial light can even impede the insects' production of pheromones, scent secretions that they send out to communicate with and find other members of their species. The darkness at twilight is the signal for the hormonal system to activate. Lights turn mating impulses off, and the night's scent trails fail to appear.

One species afflicted by this is the cabbage moth, a big, speckled moth that is found throughout great parts of Europe and Asia. The adult cabbage moth crawls out of its chrysalis in May or June, and within minutes it begins to search for a partner. The female takes the first step by extending her antennae forward, flapping her wings, and secreting scents, at around ten in the evening. A male interprets the scent and does the same, pulls back his antennae, makes several rapid wing movements, and sets out to find the good-smelling female. When they meet, the male brushes against the female's body with its antennae to feel if he has found the right mate. Then he resumes his rapid wing-flapping, and the mating is initiated. The two spend the night together, one wing of the female around the male's body, then she leaves to lay the fertilized eggs.

The entire mating ritual occurs in the dark. In laboratories, the female emits fewer pheromones in the presence of artificial light, and furthermore, the composition of the scent is completely different from that emitted in darkness. So mating never gets started. The females wait in vain in the darkness, the males wait in vain for the right scent. Any larvae that are produced and eventually pupated are also at risk of hatching prematurely. Darkness is crucial for the pupae. Long nights help maintain their rest, while light causes them to transform into moths all too soon, and they can also be hatched in the fall or winter, when no food is accessible. Light, or the absence of darkness, betrays the insects in all their stages and leads them to death.

Mass Die-Off

In addition to the previously mentioned German study, there are other indicators of alarming developments in the world's insect population. Anyone who, like me, is old enough to have driven a car during the last century would have found dead insects on the front bumper and lights. Studies from then indicate that hundreds of billions of insects may have died each year on speeding cars with illuminated headlights in the dark of night. Anyone still driving a car now can, however, attest that the problem has decreased—not as many insects get stuck on the car. That's called the windshield phenomenon, and it's a tangible, if anecdotal, observation about the number of insects near our roads. We see it in Sweden, in England, on the Continent, in America, and in the tropics—everywhere. Often it's amateur entomologists, butterfly collectors, and field biologists on excursions who've accumulated this type of data, and as yet few long-term studies have been done. But a Danish researcher, Anders Pape Møller, put his own car and its windshield to use in measuring the change in the number of insects. For twenty years, he repeatedly drove the same distance and found a significant decrease over time in the quantities of insects his car killed, which also corresponds to a decrease in the number of insect-eating birds during the same period. Møller thus showed that the windshield phenomenon is real, adding an incentive to take the problem seriously.

Much more research has been done on how to kill insects than on how to save them, which is fairly telling about how we humans operate. For example, there are well-documented tests on how to overcome pests with light traps.

The measuring of the biomass of insects in Germany wasn't initiated by the state or a university, either. It was the members of the Entomologischer Verein Krefeld, the Krefeld Insect Society, who began this substantial work. The association, formed in 1905, and has acted in the interests of insects for over a century. Fifty members, with the collection of nearly a million insects between them, crowded into their meeting room in an old school building in central Krefeld, North Rhine–Westphalia. Ten times more insects were to be found in labeled jars, stacked in school classrooms like a chaotic museum, and the collections now have cultural-historical protection. The members are not educated zoologists. There are instead priests, publishers, technicians, and teachers, all of whom, however, are prominent in their respective areas. One of the association's more renowned members, Siegfried Cymorek (1927–87), for example, was awarded an honorary doctorate in Zurich, even though he never graduated from primary school. The association has published over two thousand articles on insects, taxonomy, and ecology, and today—as a result of their sounding the alarm in 2013 about insect die-off—the research world has woken up and supports the association in their project. The interest in insect studies has also exploded at universities worldwide.

Life on earth has collapsed five times. The last time was 65 million years ago, when the dinosaurs disappeared along with three-quarters of the rest of the animal population. Today about 40 percent of all insect species are threatened with extinction, and an Australian-Chinese compilation of the collected world insect data from 2019 shows that we're moving into the earth's sixth mass extinction. And humanity is the cause. Of all the different insect groups studied, things have been the worst for different varieties of moths. Among these, more than two-thirds of species are in decline and a third are acutely endangered. The authors of the article note that the number of insect species decreases by almost 3 percent annually, which, if that pace continues, would imply that in a hundred years there'd be barely any insects at all, greatly imperiling the earth's ecosystem.

A few years ago, pictures from orchards in China's Sichuan province showed thousands of workers, each equipped with a brush as they climbed trees to hand-pollinate blossoms, work that would usually have been performed by bees. A fast worker could pollinate about ten trees a day, whereas a small colony of bees can handle a hundred times more. We don't know if our employment offices will in the future be looking for pollinators for our crops, but with fewer and fewer insects in the wild, mankind will inevitably be impacted.

The insect die-off discovered by the members of Krefeld's insect association was initially not associated with light pollution, despite the area's proximity to the industrial cities of the Ruhr and to Europe's most densely populated country, the Netherlands. This was partly because many of the affected insect groups are diurnal. But while the German analyses were in full swing and the decrease in flower pollination was evident, some Dutch researchers observed in the journal *Global Change Biology* that moths, which fly in the dark, seem to have decreased more than other insect groups. And the decrease was particularly striking in urban environments. That light is an important piece of the puzzle began to be clear.

Ecologists from the United States and Canada soon began collaborating with their colleagues from Australia and New Zealand. They were all convinced that light played a greater role than had yet been demonstrated. After all, the circadian cycle would normally have been the most constant thing in the whole ecosystem. That is, until we started using artificial light. Working together to get an overview of the situation, the research team gathered all the existing research about insects and light. They found about a hundred scientific articles claiming that artificial light adversely affects insects.

Artificial light can prolong or shorten reproductive cycles, induce hatching, and affect metamorphosis, the insect's transformation from larva to pupa to adult. Light can change the conditions for hunting and pollination, affect food intake, flight, and migration—in short, all the stages of an insect's life.

If we turn back the clock, to the beginning of the twenty-first century, the phrase *light pollution* was virtually unknown. That is, for anyone other than astronomers. There were occasional studies on how light affected birds and turtles, but not much more—not even bat researchers were talking about the impact of light on bats. Yet even now, we're just at the beginning phase. We still know too little about how light and dark affect our ecosystems. Too few experiments and studies have been done, on too few groups of organisms. A good example are spiders, which are often active in the evening and at night. Spiders can easily be awakened or made inactive through the use of light. They are never jet-lagged or sleepy; it's just off or on. So they're a perfect study group for light pollution and are high on the list of animals that the world's relatively few circadian (light-dark) biologists would like to study.

A whole world out there is governed by small changes in natural light, ecosystems that wake up and are set off by different times and programmed by different light intensities and wavelengths. One animal falls asleep, another animal's work begins, and chains of events, hormone cycles, and behaviors begin and end when the light shows, in what for humans are sometimes subtle ways, exactly what time of day it is.

With increased knowledge also come improved possibilities for solutions. The more the attention on the impact of light on ecological systems and our own well-being, the closer we'll get to reconciling society's need for light with nature's need for darkness.

PART II

The Night as an Ecological Niche

See in the Dark

The first eyes saw the world 540 million years ago. The Ancient Times came to an end and were replaced by the period known as the Phanerozoic, which, loosely translated from Greek, means "visible animals." And that's exactly what that era was about.

Everything happened during a period of 10 million years. Suddenly the animals were simply there—and lots of them. It's called the Cambrian explosion. For a long time it was thought that this period saw the origin of the multicellular animals because fossils found in mountains all over the earth show such a clear boundary between the Precambrian and Cambrian, from almost no animals at all to a huge abundance of animals over a geologic night. However, the multicellular animals had existed long before that. There have been many theories about the background to the Cambrian explosion, but what many people today consider to be the most important explanation is the occurrence of predation; that is, animals began to eat each other. The genetics were there, the individual life-forms were there, but the physical variations between them were few. All individuals resembled one another and did the same things in similar ways, at least until the threat of being eaten started to loom over them. The Cambrian is like the big bang for visible life. The building blocks for the multiplicity of life had been around for a long time, but animals began to take form in earnest after the explosion.

Animals began to be able to move. They also developed protective spikes and hard shells, shells that have been fossilized for us to contemplate half a billion years later. But perhaps even more important, the evolution of sensory organs took off, as did the arms race between predators and prey. To be able to hear, smell, feel vibrations, and see where other individuals are gives a head start in the struggle for survival and fertilization. The contest between predators and prey has been evolution's strongest driving force ever since.

The first real eyes of life also now took shape. The ability to see conveyed a huge advantage over competitors—eyes with which to see prey, eyes through which to escape other animals on the hunt. Although the ability to respond to light had existed, only suddenly now did some kind of image vision develop. Previously, the various life-forms had more or less hovered in darkness, hidden from one another. But in the Cambrian, animals could filter sunlight through lenses onto their retinas. We find traces of this development in rock deposits from China, North America, and also in Sweden.

I camped for a summer on the slope of a quarry in southern Sweden in the county of Västergötland, among the mounds of petroleum-scented stone slabs that stuck up like charred history

books in black clay slate. I was looking for fossils below the vertical walls of the quarry, which effectively concealed the light
from nearby villages and farms. The summer night's reflection
on Lake Vänern provided the needed light for my evening tasks.
Here, where sandstone turns into alum shale, the geological story
of Sweden's journey from the tropics to today's northern position
was evident. The journey took half a billion years, and many
animals were fossilized in the mountains along the way. The
black alum slate contains snapshots from life in the Cambrian.
The fossils look like hieroglyphs in the blackness, sometimes rust
red from oxidation or slabs of shimmering green slate.

The slag heaps around the quarry and other nearby alum
shale quarries are called *rödfyr* and are remnants from a bygone
era when alum shale was burned in large kilns. The high organic
content of undecayed animals makes the stone efficient as a fuel,
and these kinds of lime kilns still stand in many quarries. In
addition to energy, the shale also yielded alum, a salt that has
been used medicinally, among other things as an anticoagulant,
but more than anything else, it's a dye for textiles.

Two hundred years earlier, roughly thirty miles from my slate
mounds in the quarry, a former blacksmith's son, Sven Malmberg, and a woman called Stina Andersdotter were employed
as a servant and a maid at an alum quarry. It was one of the
oldest of its kind in the region, founded by Baron Johan von
Mentzer (1671–1747) in the eighteenth century. The quarry is
located in Dimbo parish, whose main town, also called Dimbo,
is referenced as early as the 1100s. The place name comes from
the Old Swedish word *dimber*, which means "mist, hard to see,

fog." A brook that used to be called the Dimma might have contributed to the haze over the area. The fog, the smoke from the quarry, and the petroleum-smelling slate must have lent a surreal look, perhaps as if the Gjöll itself, the underground river in Norse mythology, ran through here.

The work was hard and dangerous. Uranium and arsenic seeped out of the slate and contaminated the water in the vicinity. But Sven Malmberg and Stina Andersdotter found each other in that environment and quickly became a couple. Within a short time they'd have a son, Johannes Svensson, and a grandson, Johan, who was my grandfather's father and my namesake. I'll have cause to return to him later in the book.

Contained in the black slate are pieces of lime that smell of petroleum, called *orsten* or "stink stone" due to their unmistakable fragrance. *Orsten* consists entirely of the shells and fossilized remains of 500-million-year-old small creatures, not large but so well preserved that the area of the Västergötland *orsten* has become a *Lagerstätte*, paleontology's equivalent to a World Heritage Site. If you break open *orsten*, a 500-million-year-old cemetery is exposed and a huge number of fossilized animals appear.

Among other things, you can see different species of the now-extinct trilobites, and also various kinds of crustaceans in more or less similar shape to those we see in our oceans today. There are alternately adults—fully developed individuals—and larval forms in different stages. And it's not just the shells but the bones, antennae, parts of the mouth—and eyes. Under a microscope, the very first advanced visual organs on the earth are revealed, and not just from one group of animals, but in a number of different kinds of animals with varying family histories. In the *orsten*, you can chart the development of the eyes' ability to capture light, from the simple nauplius eyes of the larval stages to the complex eyes of the adults, the same type of eyes that we find in insects and crustaceans today. Every nuance and every single lens is preserved, and it's clear that during the Cambrian period, vision had already nearly reached today's advanced levels.

The Eye

Our human eye consists of a round, jellylike vitreum. At one end a lens is suspended in muscle fibers, shaped to refract the rays of light and capture the surrounding reflection of photons both close up and on the horizon. Under the protection of the cornea lies the iris, named after the rainbow goddess of the same name. The iris controls the window to the world, that is, the pupil, the size of which varies with the amount of light. In the darkness of the night, the pupil dilates to receive all available light, while it grows smaller in daylight. The iris not only controls the amount of light moving toward the retina and optic nerve, it also gives information to other people through its colors. The impression we make on others lies to a not-insignificant degree in our eyes' unique color combination, in their movements, and in our exchanges of glances. The goddess Iris was the messenger between the gods and humanity, between the heavens and the earth. In the same way, our iris helps us to interpret the world around us and to convey our reactions to it.

When the light reaches the back portion of the vitreum, it's captured by the retina, where it pauses for a moment and is changed into electrical signals. The optic nerve sends a steady stream of data to the brain to interpret. The light is turned into images in our interior, images that we can summon even when we blink, can come to life in dreams, and be communicated through words. Through our eyes, we've captured our surroundings.

On the retina we find cones, the color-sensitive eye cells that sense either blue, red, or green, which in combination create all our perceived nuances. The more cones, the sharper the image. The three different types of cones we have give us access to all the colors of the rainbow. But not to all the light. Birds have eyes that see more wavelengths than that, and just like we humans, they are visually directed animals. They rely almost completely on their good vision. But while we have to make do with three types of cones (three colors), the birds have four. In addition to our three cones, they also have cones that respond to ultraviolet light. Plus, birds have oil droplets in their retinas. These work much like a camera filter on your smartphone, allowing shades of colors to be perceived even better. So birds experience the world in a slightly different way from us—it could be argued that they experience a little more of it. But compared to the mantis shrimp, both birds and humans are put to shame. Mantis shrimp have up to sixteen types of cones, and their experience of color is completely beyond our comprehension.

Human eyes are largely adapted for daytime, and with the onset of darkness and the decrease in the bombardment of photons through our growing pupils, our retinas change shape. Our cones, which need light to give us definition and color sensitivity, lose their ability to capture certain light wavelengths. Details become blurred and colors fade. If people have patience on a dark evening without too much unnatural light, they can experience how their vision gets better and better. When dusk drains the world of color, when there's not enough light for the cones, then the work is taken over by the rods, the photoreceptor cells on the outer part of the retina. Rods are very light sensitive but, unlike cones, give us no information about color or wavelength. As the rods are activated, we see more and more details in the night, but we do so in gray scale. The chemical capture of light has a long history, much longer than that of the eye, with its ability to see objects. We can trace the capture to earlier than the Cambrian explosion, to dinoflagellates, single-celled algae that produce the protein rhodopsin. Translated from the Greek, *rhodopsin* roughly means "violet sight." Rhodopsin is the light-sensitive substance that we find in the rods of the retina that allows us to see in dim light, although not enabling us to distinguish between different colors. The gene that contributes sight to us is found, strangely enough, in an organism that doesn't even belong to the animal kingdom.

Unlike humans, many animals are more adaptable to the light of night than to the shades of day, including most mammals, whose visual sense has evolved for a life at dusk. Many have only two different types of cones but have adaptations to facilitate the capture of light by the rods. An example of such an enhancer is the tapetum lucidum—an extra membrane that allows twilight rays to pass over the retina twice, for maximum harnessing of light. Who hasn't at some point seen a pair of cat eyes shining in the night? What makes those eyes appear to shine is their tapetum lucidum.

In the sprawling bat family tree, the megabat has its own branch. This large, spectacular bat can be seen flying at sunset in Africa, Asia, and Oceania and lives on fruit or nectar. Unlike their smaller relatives, these bats use night vision rather than echolocation to orient themselves. Fruit bats eyes have their own unique adaptation. They have lots of veins covering the fundus behind the retina. These blood vessels passively and continuously supply the retina nutrition and oxygen without the retina itself needing any bulky blood vessels, which would obscure and shade the rods. Before the discovery of echolocation, most people thought that bats had almost supernatural night vision. That's why there are so many stories about how bats can see in the blackest of nights, aided by their blood. In chronicles of folklore from southern Sweden dated 1874 it says, "Coat the eyes with as much blood from a bat as possible and you'll see as well at night as in the day."

Bats' blood was also alleged to cure eye injuries, or to endow one's vision with philosophical clarity. That belief wasn't simply folklore, but was also held by learned people, such as Albertus Magnus or Albert the Great (circa 1200–1280). A medieval theologian and naturalist, he would be canonized. Albertus Magnus was one of the first to interpret Aristotle's texts and had a great knowledge of the teachings of antiquity. He was said to be fully convinced of the power of bats, and that he rubbed his face each night with bat blood to endow himself with as good night vision as possible, so that he could continue reading books and parchments until well into the wee hours. Maybe he made use of this trick when he wrote his work *De animalibus* (About animals), in which both dragons and unicorns occur, in keeping with medieval practice.

Nocturnal Senses

When darkness settles over the meadows and forests, shrews and hedgehogs hunt for their food. They sniff, wait, and walk carefully forward. Mice and voles also dart past, gathering food. They are monitored by owls in the shadows, who wait for the right time to strike at their prey. Their silent wings give an almost eerie impression out in the wilderness at night. Of the approximately two hundred owl species in the world, the vast majority are loyal to the dark. When the usually solitary owls nest in the early spring, you can hear their characteristic hooting reclaiming the sunset or the coolest hours just before dawn. The mountain owl's desolate song sounds across the plains in the early evening, while the horned owl's fast, repetitive sounds belong to midnight. The female and the male stay together while the owlets mature, then go on different paths and return to their solitary lives.

In agrarian society, the owl was the friend of the farm. Each night, along with the farm cat, it kept away voles, mice, and young rats from cereals and food stores. The rodents would otherwise multiply rapidly. The owl is often associated with mysticism and wisdom because it can find its way in the dark. Athena, the goddess of wisdom in Greek mythology, has an owl for a companion. But the owl's connection with darkness and all the unknown that it contains means that it's also surrounded by superstition. In earlier times the tawny owl's plaintive cries in particular were interpreted as a bad omen and a harbinger of death.

The owl's large eyes have a hard time detecting details at close range. But they effectively capture the weakest of stray twilight rays on their retinas, which are packed with photosensitive rods. At night the owl can see a hundred times better than we humans. Its head can be turned three-quarters of a turn when it scouts and listens for clues, and the most nocturnal owls, such as the barn owl, have their whole head shaped like a satellite dish to be able to pick up even the faintest sounds from prey. Because their right and left ears are differently shaped, owls clearly know which direction sounds are coming from. A delay of a microsecond of the sound between right and left ear is enough for an owl to hear the slightest rustle and locate it with the utmost accuracy.

Another nocturnal bird that has inspired myths is the European nightjar. Its drawn-out buzzing, jarring sound is a classic component of the Swedish summer night. From a distance, it may sound like a passing vehicle, but when you get closer, you can identify the individual tones in the strange sound. If you catch the eye of the European nightjar in a light in the dark, it returns the gaze with shiny large ruby-red eyes. Its tapetum lucidum, the reflective membrane, makes its night vision effective. The nightjar's genus name, *Caprimulgus*, means "goat milker," which alludes to an old belief that nightjars milk goats, harming them and eventually making them go blind. Maybe it was thought that nightjars needed to rob other animals of their sight to enable themselves to see in the night?

In the autumn the nightjars migrate farther south, just like so many other Nordic summer guests. And it follows a strict lunar calendar. When the full moon shines, it stops to eat because the nocturnal light facilitates the capture of insects. As the moonlight wanes, the nightjar flies ever-longer distances between its breaks.

In Sri Lanka, the owl is said to be the bat's consort, and in an older Swedish dialect the nightjar and the bat can share the same name. All of them are animals that fly in the night, and up until the 1600s, bats were still considered birds. But while the nightjar relies on its good eyesight, and the owl on both sight and hearing, the bat has a sixth sense to find its way in the dark: echolocation.

In the eighteenth century, several zoologists were convinced that bats had particularly sensitive skin on their wings that allowed them to navigate in the dark, that they simply *felt* their way forward. But in 1793, the Italian Lazzaro Spallanzani (1729–99) along with the Swiss Charles Jurine (1751–1819) showed that bats' ears were secret navigation instruments. By putting blindfolds and earplugs on different bats, they ascertained that only those blindfolded could fly through an obstacle course in darkness. The blinded bats could also capture insects, unlike those deprived of hearing. Records of Spallanzani's and Jurine's experiments are preserved in a number of letters that they exchanged, but Spallanzani's dissertation on bats, titled "Trattato dé Pipistrelli," was for unknown reasons never completed for publication. He was, however, widely known in scientific circles for his meticulous observations and studies of, among other things, the nightly movements of eels. Spallanzani is said to have given E. T. A. Hoffmann (1776–1822) the inspiration for his short horror story "The Sandman," about a scientist who constructs a female robot that drives a young man to madness.

Not until 1938 did researchers confirm that bats navigate through their hearing and that Spallanzani and Jurine were right. Harvard student Donald Griffin (1915–2003) had read a theory that bats used high-frequency calls, sounds beyond human perception. These ideas had been put forward after World War I, when the defense industry was experimenting with sonar technology, the art of interpreting sound echoes. But the theory had never been tested. Griffin contacted a physicist who had designed an ultrasonic detector, an apparatus to detect sounds above the frequency humans can hear.

Griffin took a colleague and some bats with him into the physics lab. Soon, for the first time, they could hear the sound of the bats and together formulated the concept of echolocation.

Griffin died in 2003 and was active to the end. The previous year he'd been at the North American Bat Conference, which is held every year in the days around Halloween. That year it was held in Burlington, Vermont, and the auditorium was filled to capacity when Griffin took the lectern. In bat circles, he's like Charles Darwin himself. After his lecture, it was hard to imagine that anyone would want to listen to me talk about bats' vision, a fifteen-minute summary of my upcoming dissertation (echo-location and all myths to the contrary, bats are not blind!). But Griffin sat in the front row and showed how curious he still was about bats, sixty-four years after his groundbreaking discovery.

Biosonar or echolocation means that with the help of sound and its echoes, different objects can be detected and located. Sound waves that bounce against an object and return to the sender can provide information regarding an object's distance, size, movement, and speed. The shorter the wavelengths—the lighter the tone used—the fewer objects the sound waves can reverberate against. We humans can use our voices to create an echo in a valley, but a bat can get tiny insects to reflect sound. Thus the bats can catch their prey in total darkness.

Using sound in navigating and searching for food, in com-bination with flight, is the recipe for bats' success as a species. But they aren't alone in using sound. Whales also use the same technique, albeit using lower frequencies, and so do the cave-dwelling oilbirds (guacharos). So do shrews, whose ears perceive ultrasound just like bats, and thus all these animals can navigate in the dark, whether it's in murky water, deep forests, cave systems, or underneath the night sky. Echolocation and the ability to fly developed in parallel in the early bats in the prehistoric forests, about 60 to 70 million years ago. Eventually bats became experts at hunting nocturnal insects, as at night the bats could avoid speedy predatory birds. Nighttime and darkness became an ecological niche in which the bats found refuge from their natural enemies.

Twilight Animals

When reason sleeps, the monster awakens. Francisco de Goya's (1746–1828) famous work depicting a sleeping man bent over a table reveals the night to be a foreign world. Behind him looms the darkness and dreams in the form of cats, owls, and bats—the night's beloved fantasy, inspiring and feared animals.

But human beings aren't adapted for activities in the dark. Our senses are simply too constrained, and, moreover, our brains require a lot of sleep to sort impressions and events. When the darkness of night falls, we usually become sleepy and withdraw. When we save our Excel files in the late afternoon and turn off the lights in the office, or maybe even more commonly just leave without turning them off, we begin a shift change. The traffic jams are slowly rolling out of town, work vehicles are silent, and the sun's rays are changing color to red. When the streetlights blink on, we know that it's evening and soon time for dinner, maybe a fair bit of time on the sofa, and after that sleep. Night workers start their day, some nighttime hikers start moving, but in large part people make a lot less fuss in the darkness.

Humans share their love of being active during the day with the majority of primates, but otherwise we're fairly unique among mammals. Most of the world's nearly six thousand species of mammals prefer the hours at dawn and dusk, or even the night, and have done so ever since the continents of the earth were still together in a single large body of land and Europe was covered in a tropical forest. While the dinosaurs were still around, early mammals moved under the protection of darkness, with well-developed night vision, whiskers to feel around with, ears alert for the slightest rustle, and a sense of smell adapted to the scents of the night. Animals that sought refuge in darkness could escape predators and had the insects of the night to eat.

Many mammals were initially small and lived in trees, shrubs, or on the ground under the protection of fallen leaves. They sustained themselves mainly on seeds, insects, and other small animals. Only when the sun was setting did they venture out in search of food. From these ur-mammals evolved all the genetic basis for everything from carnivorans to whales to bats, all the fauna we see around us today. And they were all born in the shadows. When the dinosaurs eventually died out, the mammals took over the vacant niches, spread out, grew, and even ventured into the daylight. But our dogs still find scent traces more easily in the evening moisture, and it's during my twilight trips in search of bats that I see moose crossing the roads, marten eyes glistening from the edge of the forest, and foxes hunting in the fields. Mammals are still largely twilight animals that are least active in the middle of the day. Even animals that we associate with daytime activity, such as certain squirrels, ungulates, and most predators, have some of the characteristic night senses left.

Cat owners are not infrequently woken up by their cat's meowing and wanting to go out in the middle of the night. At home, the cat may be timid and lazy, but outdoors it's still wild. In 2013, the BBC documentary *The Secret Life of Cats* was broadcast. The documentary was based on studies of fifty cats in a residential area in the UK. Each cat was equipped with a GPS and microcamera, and every movement was documented. All cats were found to have well-defined ranging areas, and of a smaller size than previously thought. Several cats might divide a given area, and they patrolled at different times, some at night, some during the day. Some cats were decidedly twilight animals; others made use of the entire day. Cats would commonly enter other cats' homes and eat their food. When the thieves were discovered, a power struggle would ensue between the cats.

With the help of modern night cameras that can operate in low light, in recent years we've seen that more animals are active at night than we previously knew. Cheetahs, for example, which normally hunt during the day, use the full moon on the plains to take almost as much prey in moonlight as they do in daylight.

Elephants have always been most active during daylight on the savanna, even though matriarchs sometimes lead their herds to water holes when the night's beneficial coolness has descended. Moving in the dark gives nocturnal hunters such as lions and hyenas the advantage, but despite this, several African elephant herds have begun to show more and more activity at dusk in recent years. The reason is poachers. The darkness of night protects the elephants from poachers, and it has been shown that the oldest members of the herd are well aware of the reserve boundaries, where they can go safely and where they are exposed to stalking by humans. Only under cover of darkness do the elephants move between two nature reserves.

In desert areas around the world, the sun's rays are so hot that they kill, which is why night is a sanctuary for all living things, including cacti that bloom only in the dark, in anticipation of the bats' pollination trips. Mongooses are catlike predators that live on the savanna and are close relatives to the meerkat. Meerkats became famous through the Disney character of Timon in the film *The Lion King*. Mongooses have always been considered a daytime mammal, but two researchers looked more closely at what they do at night. Not surprisingly, they turned out to also have a relatively active night life. Several individuals changed their residence, one chased away an intruder, another examined holes in a different area. The mongooses didn't lie down and sleep through the night; they had interactions both inside their dug-out cavities and out in the night air.

Life has evolved in accordance with the day's alternating light and darkness, and the more animals we study, the more we realize that both day and night are equally important for their ecology. In an increasingly illuminated world, the boundaries of the different times of the day are blurred and activity patterns change. We still know little about what this means for animal and plant life.

Sing in the Wrong Light

One of the most familiar sounds of the night, especially at southern latitudes, is the persistent, high-frequency chirping of crickets and frogs. Many people find the sounds soothing; there are recordings of the sound of crickets for people who want to shut out other sounds and possibly fall asleep more easily. But the frequency of the sound of crickets can be right on the edge of what humans can perceive, and sometimes those sounds can be extremely unpleasant to us.

The crickets use sound both for marking territory and as a mating song. The male most often plays his serenades in the evening. The cricket's song is quite strongly linked to the transition between light and darkness, and small changes in the intensity of the light are enough for the cricket to put off its song and miss the mating ritual. In addition, if the light is too bright, the growth of the young crickets, the nymphs, is delayed. The length of the day tells insects what time of year it is, so that they can speed up or slow down their growth based on the available light.

The songs of the crickets attract attention not only from the opposite sex, but also from predators. Crickets singing in the wrong light run an extrahigh risk of becoming food before they've had time to mate, a problem that crickets share with many insects. Moths hanging out around lights and light posts after dawn are easy prey. Birds, as well as rats, lizards, toads, and spiders, have been shown to take advantage of insects' gathering and remaining at lights. Predators have a buffet laid out for them. Even while it's still dark and the moths are fluttering around in the streetlights, they're easy targets. At that time, bats are the enemy more than anything else.

My colleagues in the small bat group at the University of Gothenburg studied interactions between moths and bats in Gothenburg at the turn of the twentieth century. The researchers used a small ultrasonic generator to emit high-frequency sounds just like a bat makes. With the generator, you could make moths react in different ways. As though following signals from a remote control, a selected moth was made to dive to the ground, turn over a small coin, or do a loop out in the night's darkness. One push of a button and its wings stopped beating; another, and it turned off course to confuse any pursuer. When a bat sends out its echolocation sounds in search of prey, it can do so up to 130 decibels. So a prey animal that hears high frequencies has no problem picking up the sounds of bats. Having developed just such a defense, moths can normally avoid ending up being sitting ducks for hungry hunters at night.

When the insect sonic generator was tested on moths under streetlamps, however, the moths were seemingly deaf and had no reactions whatsoever. It seemed as if light misled their auditory system. But moths react differently to dangers depending on whether it's light or dark. Birds are the main threat during the day and bats at night. The results, which were first published in the scientific journal *Animal Behavior* in 1998, have been cited many times in recent years, for they fit perfectly into the new field of light pollution.

The eyes of nocturnal insects are particularly sensitive to green, blue, and ultraviolet, and this kind of light more than anything else motivates them to fly toward lights. So orange streetlamps, which are the most common along our roads, attract fewer insects than the older white lamps (mercury lamps), which all emitted a large amount of light in shades of violet. The clouds of insects that could previously be collected from about fifty yards around each light post are now smaller. But in a trade-off, there are more lights.

Moreover, today the plethora of different kinds of lights, all with different wavelengths, have different allures for insects. A standard streetlight attracts insects from about twenty yards, but sometimes it's up to fifty yards. Given that streetlights are usually closer together than that, it's extremely difficult for insects to cross a road without being ensnared by a light source. This means that every road, even the smallest sidewalk, acts as a kind of a barrier.

Even if we stuck to amber or even red lights, which insects to some extent ignore, or if we designed lights with well-defined spectrums so as not to disturb the insect orientation, the intensive use of artificial light means that we're always spilling light into the sky. A light without a shade radiates more light and energy into the atmosphere than down to the walkway it was intended to illuminate. This waste becomes a screen of diffuse light, so-called sky glow. On a cloudy evening when the light is reflected back to earth, the sky glow can cover our cities with a yellow dome, completely obscuring the night sky, and in that light insects have other problems.

The light of the full moon can be bright enough to keep certain insect species on the ground, as they could otherwise be easily taken by predators. They will wait for darker nights rather than fly in the moonlight and risk their lives. Normally, this isn't a problem. As is well known, the moon travels its silent path, and soon the darkness of the sky is back. The phases of the moon are predictable and recurring, constituting one of the cycles in nature that can affect and calibrate the biological clock. Sometimes, however, a cloud dimming a partial moon is enough for the insects to take wing into the night. But under a constantly artificially lit sky, even if it seems dark to our eyes, it's as if the full moon is always present and is often even brighter than that. Under such a sky, not even the clouds are helpful; quite the contrary, as these reflect the light back toward the earth.

Nature's Own Lanterns

Patterns and warning colors that on animals and insects are meant to camouflage, scare, deceive, entice, or communicate that its bearers are toxic were developed over millions of years to be most effective at a particular time of day. When we turn on a source of light, colors are reflected differently, patterns become distorted, and structures are blurred. All our different types of lights—from decorative garden lanterns to facade lighting— glow in different ways, with different nuances and wavelengths. They all cause different kinds of shine on whatever they illuminate in the night. While color contrasts are enhanced in one kind of light, they decrease in another. Instead of being hidden in the dark, a creature could suddenly be fully visible. Instead of attracting a partner, you could be discovered by a predator.

Some kinds of moths do everything they can to be seen at twilight, and this gives rise to a recurring grand spectacle each year. Sweden's ghost moths are large and chalk white and cannot hear. They fly over unmowed meadows in the pale, washed-out darkness of a June night, just as the fog drifts off streams and over cairns, and the woodcock takes its last tour of the evening. In the meadows the moths quietly start using what looks like an invisible elevator. Up and down, up and down, in a tireless rhythmic movement. Before you know it, the meadow is a sea of bouncing silver-white moths. They're all males, who've show up in hopes of finding a female. Their mating dance takes place in the same place, at the same time every year, and their white wings are clearly visible against the dark green grass in the dwindling evening light.

But the ghost moths are playing a dangerous game. Soon a northern bat dives down from its patrolling height of about four yards above the ground, makes an elegant U-turn, and catches a moth just above the top of the grass. Several times I've witnessed both the ghost moths' flirtatious dance to ensure the next generation, and how the northern bats materialize from the shadows behind the tree curtains to immediately plunge down into the swaying grass. After about forty-five minutes, everything suddenly stops. The meadow's silver-white wave subsides, the bats move on, and an owl hoots out that the evening's show is over. Summer twilight has passed and, with it, the shortened night. Dawn is rising again.

In recent years, I witnessed a lone ghost moth in an overgrown ditch as he rode his invisible elevator in the summer night. Otherwise, it's been a long time since I've seen the moths dancing over the countryside in the evening. Twilight is chased off by artificial light, large-scale agriculture is taking over, and the old meadows have become grown over. The moth populations are dwindling.

We know that the ghost moths fly at a time when the natural light best reflects off their white wings; that's how the males gain the females' attention. In artificial light, the contrast between the light color of their wings and the evening sky is masked, and the mating dance literally fades out. No one has done a formal study on the effects of this, not on the ghost moths in Sweden. But other species have been seen to be affected in a similar way, species that also turn up in the dark with spectacular signals. A clear example are the glowworms, those beetles that have built-in lanterns and glow with a bright yellow-green shine in summer meadows and in clearings in places such as Sweden. The females lack cover wings and look a bit like a caterpillar, which is why she is incorrectly called a "worm." At dusk she climbs out on a leaf or blade of grass, directs her glowing backside toward the sky, and waits. Soon males come flying, lured by the seductive light, which looks like a small diode in the night. The males also glow, but only mildly, not as phosphorescent green as the females.

In some places, glowworms are present in such large numbers that they've become tourist attractions. One such place is Waitomo in New Zealand, where half a million visitors descend each year into a bustling labyrinthine cave, greeted by a ghostly glow from above. Hundreds of thousands of glowworms give off a bluish light, like the glitter of the moon across a rippling sea. The glowworms are of a different variety from the one in Nordic countries; these in Waitomo are mosquito larvae, which, after they have hatched on the edge of spring, spin a small nest of silk thread. From each of the thousands of these nests about twenty long threads hang, each outfitted with sticky drops reflecting light in all directions and creating a carpet of pulsating mystery under the ceiling of the cave. The threads catch gnats and other small insects, attracted by the larvae's built-in diodes. The larvae are sometimes called spider worms because they spin catch threads and trap prey in the same way that spiders do. The adult mosquitoes also glow. Why isn't known for certain, but it could have to do with mating. This is the only task the adults have. When they have found a partner and the female has placed her eggs on the cave walls, they die, making room for a new generation of luminescent mosquito larvae.

Another cave in Tasmania has glowing creatures of the same genus. They shine just as seductively and attract visitors from near and far. The light of the larvae seems dimmer and faded in brighter conditions, so the tourism industry does its best to completely avoid artificial light. In addition, the display is best experienced with eyes that are adapted to the darkness as much as possible. I have walked in a New Zealand forest before dawn, guided only by the light of thousands of glowworms in the bushes, as though in the wake of a fire-breathing dragon. They gave off so much light that not once in the three-mile-long promenade did I have to turn on my flashlight. But in a commercial context, lighting is still needed. In the Tasmanian cave, the effects of the lights used to help tourists see where they set foot have been studied. It seems that the insects handle it well. They recover quickly and their natural rhythm continues normally, as long as the lights aren't on for too long or too often. But that's the case for mosquito larvae in caves, where it gets really dark, where the effect of a light disappears quickly. It's worse for the lightning bugs in Europe, the beetles that crawl up on grass blades in the open air. Their light is turned on after sundown and is strongest and most effective just before the first hour of dawn, while the darkness is still dense.

In England, the public has been reporting discoveries of glow-worms since 1990, at a time when it was feared that the dramatic creatures were disappearing. Even though the number of sites where they were found increased at first, you can now see a nega-tive trend—the glowworms *are* disappearing. The ones found now are mainly in nature reserves and in the countryside, far from the light of the cities.

A female glowworm waits out the day in a hiding place, turning on her lights shortly into the early twilight. Under an artificially lit sky, her light will stay off, despite her inner clock having aroused her desire to mate. Even if she climbs out on the blade of grass with her light on, it may not be visible. Females' signals are drowned out in all the surrounding light, and even light equivalent to that of the full moon is enough to reduce the males' perception. The males may also fly to other lights shin-ing from afar, in the hope of finding a large colony of females. The males may fly far away only to discover that they've been deceived by lights from windows, cars, or streetlamps.

Unlike the green lights of the glowworms, fireflies have a yellower light. Fireflies can be found mostly in warmer areas, they flash in yellow, amber, orange, and green, communicating with one another in what can be compared to Morse code. In Chinese mythology, fireflies appear when the grass of the plains is on fire, and in Japan they symbolize the light of human souls from the past. The firefly fascinates and attracts us, and people in several parts of the world tell stories about how children used to collect fireflies in jars to use as lights at night, but today those are mostly tales to tell the grandchildren.

Just as in the case of the glowworm, the name *firefly* is misleading. The firefly is also a beetle, and *glowworm* and *firefly* can be different words for the same beetle species, in which the female is stationary, glowing, while the male flies. When the male in the mating dance flashes his light, the female responds by flashing back. The two create an alliance and glow together in confirmation. The light can also guide the firefly in the hunt for food in the dark and scare off predators by signaling that the firefly is poisonous. In some places, entire colonies of fireflies can be seen flashing in sync, like a pulsating control panel during dark tropical nights. One such place is mangrove swamps in the Philippines, where female fireflies sit in the trees, blinking a light green, while the males fly above the trees, blinking back, all together, as if it were a surreal, extraterrestrial Christmas show. It's no wonder that myths arise about fireflies as fleeing souls and about how they can show lost travelers the way home.

Light Spring

The American author and marine biologist Rachel Carson (1907–64) is one of the environmental movement's paramount early figures. Her 1962 book, *Silent Spring*, criticized the wide use of the insecticide DDT and foresaw the insect die-off we see in the world today. Without insects, there are no birds and no birdsong. *Silent Spring* is about nothing less than the dystopian silence of an impoverished ecosystem. Sweden was the first country to ban DDT, in 1969, and the rest of the world soon followed suit. Although the spring has not gone silent, the insect-eating birds, following the same trend as their prey, are in decline. For other birds, especially omnivores that can find food in cities, things are improving.

One of those urban birds is the European blackbird, which, with its recognizable melodic trilling, has been named Sweden's national bird. It was once known as a forest bird with a preference for the deep, dark woods. It was somewhat shy and settled among thickets in humid, shady environments, singing its songs. Its dark plumage blended into the foliage of the forest as it searched for food in the soil, its exact position hidden within branches and leaves. But in tandem with emerging industrialization in the nineteenth century, the blackbird became part of the city. Fewer and fewer blackbirds bothered migrating in the winter, and those that stayed sang louder and longer into the autumn. In noisy metropolitan environments, they now also sang louder simply to be heard over the traffic.

In the medieval city of Leipzig in Germany, project researchers spent their evenings studying the behavior of the blackbird. More than anything else, they wanted to know how long after sunset the birds continued to hunt and eat, and to compare results from forest and park environments to those from the downtown city center. It turned out that the brighter it was from the lights of the city, the later in the evening the blackbirds were active. Kind of like humans. It was especially clear on the edge of spring.

Sometimes on my way home from bat studies in the field in the wee hours of the morning, I hear early-morning songbirds already up at two o'clock, long before dawn. Light equivalent to that of the full moon is enough to trigger the singing of some birds, especially in the spring when they're establishing their mating territories. A male who's out early has a greater chance of impressing females with his persistent songs.

In Munich, just over 220 miles south of Leipzig, other blackbirds have shown researchers how much light they're exposed to daily. The birds have been equipped with small devices that record this light exposure around the clock, storing the information on a chip. This has shown that urban blackbirds are, on average, exposed to just over a thousand times more light at night than forest populations. While forest blackbirds live in light conditions that roughly correspond to cloudy autumn nights, urban blackbirds always have at least a full moon's worth of light. Additionally, they experience almost an hour more equivalent of daylight.

In 2019, more blackbirds singing into autumn were reported than ever before, especially in the Stockholm area. The mild November weather, combined with the city lights, seems to have given the birds a sense of spring. The normal shadow song, the quiet growl that you can hear from them in the winter, had suddenly been replaced by a trilling mating song in the middle of the November darkness. People often react strongly to the song of the blackbird, and maybe this isn't so strange. We've been influenced by the sound since we took our first steps, and their song is so intimately associated with spring that the blackbird's spring feelings spread to us. We instinctively feel that it sounds wrong to hear blackbird song accompanied by Christmas music in the city shops.

The habits of birds in town extend both earlier and later in the seasons than do those of their cospecies counterparts in forest environments. The city birds have a longer mating season and more time to sing, socialize, and find food. In some cases, because they reach sexual maturity earlier, the city birds have almost a month longer to have their young. They might even have several broods in a single season. In addition to the increased day length—even if it's artificial—it's also warmer in the city, which often also has fewer predators. But it's not a given that the city birds feel better; the extended length of day doesn't seem to give them long-term physical benefits.

One reason is that the birds' hormonal systems are impacted. The return of sunlight in the spring is supposed to initiate the desire to mate, sing, and produce sex cells. After the long winter nights are followed by longer days with higher temperatures, the eggs are to be laid, and eventually, when the insects have come back to life and the summer smorgasbord has been set, the chicks are supposed to hatch. Particularly light-sensitive receptors in both the hypothalamus and the pineal gland in the brain help to constantly adjust the biological clock to optimize for the survival of the next generation of birds. But when the light is deceptive and obscures the arrival of spring and the length of night, birds become sexually mature at the wrong time. In the best case, it gives the birds a competitive advantage; in the worst case, the timing between males and females is off, eggs hatch before there's food to be had, and the body's systems are under stress. A study in Florida showed that sparrows with West Nile fever, a viral disease related to yellow fever and hepatitis C, were infected an average of two days longer if they were exposed to artificial light during that time. So the lighting in cities inhibited the birds' immune systems, which thus increased the risk that the virus would spread to humans.

If mating takes place in the autumn, as with bats and deer, the animals obviously don't want to have their offspring in the middle of winter. The female bat therefore stores semen until the spring weather shows that it's time for the egg to be fertilized. This also allows the mating season to be extended into the winter, energy allowing. With deer, the female's eggs are fertilized directly, but fetal development doesn't start until after the turn of the year, so that the fawn isn't born prematurely. Adaptations such as these are governed by several factors in which the length of the day is an important piece of the puzzle.

In southwestern Australia there lives a small kangaroo-like animal, the tammar wallaby. The wallaby females give birth six weeks after the summer solstice. This ensures that the female has the greatest access to food and can produce as much milk as possible when the young need it most. A change in the light gives notice, like a calendar, when the time is right to start this process. But in a population of tammar wallabies on Garden Island off the coast of Australia, the young are being born later, up to a month after the usual time. This was found to be due to proximity to a naval base that cast its spotlight over the animals' territory. The artificial light obscured the natural changes, polluting the calendar of the evening sky. Hormone systems were disrupted, delaying everything from mating to birth and growth. The wallabies weren't born until that season's food had already begun to run out. Nature was again disrupted by man-made light.

The Star Compass

In Sweden we have an incredible diversity of birds during the summer, when food is most available. But when darkness steadily returns and the temperature drops, hordes of birds head south. They travel from the uninterrupted light of the polar summer to the tropics' regular cycles of light and darkness. All over the earth, birds move between winter and summer places and between rainy and dry periods to nest where the food is. Some of the migrants, such as the arctic tern, travel twenty-five thousand miles a year, that is, a whole trip around the earth.

To find the same place year after year, birds have a whole host of tools at their disposal, such as an extraordinary memory for landmarks and the ability to navigate by the sun. They can also sense the earth's magnetic field and so determine how far north or south they are—they have an inner compass. How this works has been the subject of lengthy research. Particles of the magnetic substance magnetite have been found in birds, which could be part of the explanation, but there are many indications that they actually use their eyes. In a ring along the edge of the retina are cryptochromes—proteins that are particularly sensitive to blue light. Cryptochromes are found in both animals and plants and assist, among other things, in governing the circadian rhythm; but a type of cryptochrome protein has also been shown to respond to magnetic fields. Maybe birds and other animals, too, can simply see the earth's magnetic field.

To travel several hundreds of miles from one place and find their way back year after year, birds also use nocturnal clues. Nocturnal migrations are common, especially for small birds. It is estimated that two-thirds of all migratory birds make their long-haul flights at night. In the 1950s, the ornithologist couple Eleanor and Franz Sauer built a glass cage that they took turns sitting in during the evenings. They had also placed warblers in the cage, which during the migration period showed clear signs of wanting to flee into the night. On starry nights, the birds made outbursts in a given direction, while on cloudy evenings they were somewhat calmer, more tentative, and not as clear about where they wanted to go. To test the hypothesis that birds can navigate by the stars, the couple constructed a planetarium with each star depicted as a small point of light in the ceiling. They moved in the warblers. The planetarium's projection of the starry sky could be turned on and off and rotated. As long as everything was normal, as long as the stars shone brightly in the deep-blue-painted ceiling, all the warblers headed in the same direction every night. If the starry sky was put out, though, the flock was confused and the birds headed off in slightly different directions, if they tried at all. Several of them just sat down and groomed their feathers, apparently in anticipation of better weather. The most interesting discovery was that one could influence the birds' direction of flight simply by rotating the entire projection. The birds were obviously studying the starry sky and acting according to its pattern. The question was how.

If there's a single star that most of us living in the northern hemisphere know by sight, it's the North Star. Four hundred light-years away, it's clearly visible in the night sky in the north, as a constant guiding star in an otherwise changing sky. Around the North Star, the Big Dipper and the entire starry sky rotates slowly counterclockwise, as the night progresses. The North Star is the celestial pole, the celestial vault's equivalent to the geographical North Pole. The phenomenon has been a guide for seafarers through all ages and probably also for birds. If the latter is the case, Stephen T. Emlen wanted to find out, so he picked up where Franz and Eleanor Sauer left off. Emlen worked with indigo finches, beautiful blue cardinals that move between North America and the Caribbean every year. He created a system whereby the finches tramped around in ink, and every movement, every shot in a certain direction, made small footprints on a piece of paper. Emlen researched the birds' circadian rhythm and what they saw in the sky. By carefully blocking out selected parts of the star constellations in his false starry sky, he concluded that the North Star and the Big Dipper were the starting point for the finches' navigation.

But Emlen didn't stop there and instead let the homemade night sky rotate around another star instead. He moved the celestial pole to Betelgeuse, a bright red giant star in the constellation

Orion, known to anyone who has read Douglas Adams's classic *Hitchhiker's Guide to the Galaxy*. Betelgeuse has consumed all its hydrogen, so it glows red instead of white, and the star is expected to flare up into a supernova. It could happen today or in a thousand years, and when it does, we'll be able to see the spectacle with our naked eyes even during the day. Otherwise, it's in the December darkness that Betelgeuse can most clearly be seen, on one of Orion's shoulders. When Emlen's indigo finches saw that the sky was spinning around Betelgeuse instead of around the North Star, they aimed for it to start their journey north. Another group of finches that Emlen had tricked into thinking it was autumn tried to fly in the exact opposite direction. Emlen's and the Sauers' experiments have been repeated many times in the last fifty years in many different ways, and it seems as if birds have an innate ability to study the starry sky and recognize the pattern of the stars closest to the celestial poles. This knowledge, together with their inner compass, leads them the right way on their long nocturnal journeys across the earth.

It's long been known that night-traveling birds fly high when the moon shines in a clear sky and the stars can be counted in their thousands. On cloudy nights, when rain and fog make the birds fly lower, they risk being confused by lights and buildings, and large numbers have been recorded as colliding with masts, towers, and lighthouses around the world. In a notorious incident from 1968, no less than five thousand birds, mostly songbirds, collided with a television tower in Nashville, Tennessee. And below the lighthouse at Long Point in southern Canada, seven thousand dead or injured birds were found between 1960 and 1969. As early as 1880, a compilation of birds that died at lighthouses in North America was made. The lighthouses off the east and south coasts—in Louisiana, Florida, North and South Carolina—were the most dangerous places and coincided with some of the birds' most important migration routes.

The mesmerizing attraction of light has been used as a hunting method by people in several parts of the world. In the early twentieth century in India, reports were that birds could easily be caught if lanterns were lit during nights when the fog was dense and the wind came from the south. On the African savanna in modern times, birds are attracted in the twilight with mobile spotlights for viewing by tourists. As early as 1883, before the breakthrough of electric light, Darwin's friend and younger colleague George Romanes (whom I talked about in the chapter "The Vacuum Cleaner Effect") likened the attraction of light for birds to that for moths. He wondered if a burning candle flame, an unknown phenomenon in nature, could be mistaken for a white flower or other objects that stimulated the birds' innate curiosity. He also noted that fish were attracted to lanterns hung off boats, as fishermen have known since time immemorial. Whether insects, birds, or fish, darkness is essential to all living creatures.

The Dazzling City

On September 11 every year, the art installation *Tribute in Light* is lit in New York, where the Twin Towers once stood. The installation consists of eighty-eight spotlights that together create two shimmering blue pillars of light. On a clear night, they reach half a mile into the sky and can be seen from ten miles away, despite the big city's other glowing lights. The annual tribute to the victims of the 9/11 terrorist attacks is witnessed and appreciated by tens of thousands of people, but it also coincides with the migration of birds past the city. Because the spotlights are only lit one night each year, they are an excellent opportunity for researchers to study the effects of light. Between 2010 and 2017, the birds' reactions to and flight patterns around the slightly extraterrestrial pillars were studied, and researchers found a number of strange behaviors. Shortly after the lights were lit, large flocks of birds gathered near them and flew around in circles, cawing and singing. Just over 1 million birds are esti- mated to have assembled in total on the seven occasions that

Tribute in Light was lit during the study. However, as soon as the lights went out, the flocks dispersed. The study clearly showed that light affects and confuses birds.

As skyscrapers have grown high on the America continent, the problem of light-confused birds has only increased. The lights and the sky glow from big cities blur out the stars even on clear nights and lure birds down from their normal altitude. Given how the world's big cities are expanding and spreading their light farther and farther into the countryside, and higher up in the atmosphere, nights are increasingly rarely starry enough and suitable for navigation. Once caught in the grid of the skyscrapers, the birds get stuck in a maze of confusing lights, reflective glass, and tall obstacles. Strong lights can have a mesmerizing effect on birds, and sometimes they get caught in cones of light like confused moths. The city-center light can also resemble the light streak of the horizon, so birds retreat there or go completely off course. The light can also inhibit the birds' natural night vision, even dazzle them, so that they have to fly lower to look for landmarks.

The Great Salt Lake in Utah is North America's richest bird sanctuary. The number of birds passing through every year can be counted in the millions. Beach birds and seabirds, waders, birds just taking breaks, and nesting pairs from more than three hundred different species coexist in the salt lake and its surrounding wetlands. You could say that the Great Salt Lake, although it's a smaller inland sea rather than a lake, is North America's answer to Sweden's Lake Hornborga. Southeast of the Great Salt Lake, Salt Lake City is growing, even if it's as yet a relatively small town by North American standards—two hundred thousand inhabitants—but still it hosted the Olympic Winter Games in 2002. Once an important mining community and the Mormons' refuge from persecution, the urban region is growing rapidly along the eastern parts of the Great Salt Lake toward the towns of Ogden and Logan, and the lights from residential suburbs and city centers get reflected by the salty, nutrient-rich wetlands, coloring the sky pale orange. Salt Lake City residents are beginning to understand the seriousness of this threat to the mighty diversity of birds that live and thrive in the area. Since 2014, Tracy Aviary, a botanical garden and a center for bird research in downtown Salt Lake City, has been working to educate the public and conduct public research into light pollution. Residents of all ages have joined in to count the dead birds in the mornings in downtown Salt Lake City, where each participant follows an assigned route. The center also maps "unnecessary light," such as decorative facade lighting and lights that are badly shaded or bling bright.

Tracy Aviary has also initiated a movement for residents to sign a pledge to turn off their garden lights and dim their window lights during the birds' migration seasons, which run from March to May and from August to October. Participants fill in a form stating that they are actively counteracting light pollution. Then they get a sign to put on their house that shows three birds against a black, stylized silhouette of high-rise build-ings, with the Wasatch Range, the western edge of the greater

Rocky Mountains, towering in the background. The mighty range should remain dark at night, Tracy Aviary maintains, so that the natural-light contrast between the distant mountain peaks and the salt deposits on the fields in the valleys below can be preserved for future generations.

False Summer

One late-fall afternoon a couple of years ago I was walking through Gothenburg, Sweden. The cafés projected autumn coziness and beckoned with coffee and buns. The Halloween signs had been taken down, and here and there some people had started putting up Advent candleholders way too soon. It was tempting to grab a seat with a blanket and a hot cup and fall into temporary slumber, but I continued my walk. Most of the trees had recently shed their leaves, but here and there color-ful birches and maples were lit up by spotlights from below. In the light, the trees still had all their autumn splendor, all their clothes on. From a human perspective, it was quite beautiful. But it still felt wrong.

Plants capture sunlight with the help of chlorophyll, a pig-ment found in the chloroplasts of plant cells. The green light is reflected and gives the leaves their perceived color. Newly sprouted leaves are light green and filter spring light on its way to the ground, and everything that's budding has its start there.

Soon the leaves darken to the characteristic green hue of summer. In the autumn, when the light fades and the days are compressed, chlorophyll disappears and the leaves change color to be all shades of fire, before they eventually drop. The trees are then ready for the winter cold.

Just like warmer weather, the glow of lights can deceive trees, and the falling of the leaves is delayed, sometimes until well into late autumn. On the Continent, botanists have seen how rowan and maple trees that are next to streetlamps retain leaf cover for three whole weeks longer than their cospecies that grow in natural light, despite being in a similar climate.

At the start of spring, artificial light can accelerate the awakening of the trees, cutting short their rest, making the buds open prematurely. Larger trees such as oaks and beeches can bloom about a week earlier when exposed to street light, while smaller plants are more sensitive and can be triggered to bud even earlier in the late winter. So while the darker forests of the countryside are still asleep, the city's plant life is ready for the summer before the frost has barely let go, as if the stress of city life has sunk its claws even into the plant kingdom.

Each plant has its own time of season for when it's best to flower to maximize its number of seeds. Releasing seeds early and gaining a foothold in the field before competitors can be an advantage, but spring can be treacherous. When cold strikes, newly sprouted buds are sensitive, which is why plants need reliable guidance from the environment. One example are apple trees, whose flower buds freeze off during a spring frost, a phenomenon many growers have experienced with increasing frequency. Temperature, number of hours of sunshine, and the color nuances of light all provide clues to the right time. Global warming and artificial light together can therefore create major problems within plant ecosystems.

Robert Hunt (1807–87) was the first to study plants' reaction to different light wavelengths. He was educated in physics, chemistry, and anatomy and was a writer, poet, and visual artist. Like so many of the former pioneers within science, he was a multitasker, with one foot in art, the other in the search for knowledge. Robert Hunt's life was largely influenced by photography. The camera was new in his youth, and Hunt realized the possibilities and power of this new medium. With his knowledge of chemistry, he experimented with the development of photos, and with his knowledge of physics he studied light and its wavelengths.

In the 1840s, Hunt discovered that different parts of the light spectrum affected plants in different ways. The shorter wavelengths of sunlight, the blue and violet light, are usually the ones signaling seeds to germinate, while a redder light, with longer wavelengths, normally initiates flowering. Not until the twentieth century was it identified what within the plants controls this process, when the various plant pigments, the phytochromes, were discovered. These proteins assume different shapes depending on the ambient light. When the light changes, this process takes place at lightning speed and makes a plant perceive differences between wavelengths, the colors of light.

How the plants react depends on the context in which they live. Different lights have different properties and plants react to both intensity and color. Many modern lights (LEDs) glow with a white, almost bluish glow, resembling the morning sun's mix of short and long wavelengths. Older light bulbs, such as streetlights, are often yellow or amber and more similar to the late-afternoon sky, which is dominated by longer wavelengths. This light has sometimes been shown to inhibit, rather than to accelerate, the flowering of certain plants. In England researchers have studied the meadows in Cornwall, in particular the bird's-foot trefoil, which we recognize by the characteristically shaped yellow flowers on a midsummer meadow and heaths. Normally, they attract large numbers of aphids, but a late flowering or absence of flowering can decimate populations of aphids, which in turn affects green lacewings, shortwings, ladybugs, flowerflies, and the insects that we're used to seeing flying over the meadows and that live on flowers and aphids. The domino effect begins, and the ecosystem is disrupted.

Fruitless Night

In September of 2017 a mysterious glow was reported over Härnösand, Sweden. Had a UFO landed? The local media took photos as well as testimonials from curious residents. But it wasn't a UFO. The low clouds of the dark evening were reflecting light from a newly opened greenhouse. We humans can trick plants into serving us. Greenhouses with around-the-clock lighting do just that. The natural timing for flowering and seeding may not suit us, so we can force a desired action by providing the plants with more or less light based on our demands, not their needs

Every year I get a poinsettia from my mother. It adorns the kitchen table for a while, but after it has bloomed, no one expects it to bloom again. In Mexico, where the poinsettia actually belongs, it has a longer flowering period, which depends on the length of the day, or really, the length of the night. The poinsettia is a so-called short-day plant, needing just over half a day in uninterrupted darkness in combination with the right temperature to bloom. To sell the red poinsettia at Christmas every year, the plants must be prepared in October. In greenhouses, the light is set to trick the plants to think that they're growing under the Mexican night sky.

Nearly all plants require darkness; their light-sensitive phytochromes react to the changes between light and darkness, and length of the night largely determines whether the plant should be resting or growing. For the short-day plant, the darkness is more crucial, and it might more appropriately be called the long-night plant. The darkness needs to be consistent, creating an uninterrupted rest from the influence of light. During evolution, perfect timing was coded into these genes to give the plant maximum opportunity to spread its seeds and thus its genes to the next generation.

We humans can take advantage of this. Had I been a little more enterprising with my poinsettia after the Christmas flowering period, I could have placed it in a dark closet for thirteen hours a day, and then in a bright window for eleven. The plant would then have been triggered to bloom again and again.

The alternative to short-day growth is of course long-day growth. Some plants need a short night, or a short period of uninterrupted darkness followed by a longer bright day, which is typical of our Swedish summer. Greenhouses around the world take advantage of this, and it's not uncommon to see their lights on around the clock. Global warming and artificial light reset the internal clocks of plants, risking destroying the delicate balance between plant and pollinator, between plant and herbivore, between prey and predators. If the temperature rises just a few degrees or twilight is shifted only by a few moments, the timing between flowering and the creatures who take advantage of that flowering can be disrupted.

Today we see that more and more plants remain untouched by insects and never bloom. A group of Swiss researchers studied pollination at the foot of the Alps, in an area still relatively untouched by man, where the flowering meadows rest in night's darkness in the shadow of the mountain slopes. Every night almost three hundred different insect species visit the vast, moist grasslands to pollinate about sixty different varieties of flowers. One such plant is the cabbage thistle, whose pollen- and nectar-rich bright flowers attract insects from near and far with their attractive scents and flower bowls that reflect ultraviolet light. The researchers studied one hundred cabbage thistles, spread over ten meadows. Half of these were allowed to remain in the dark, while the others were lit up by lamps common in today's streetlights. The study showed the number of visits by insects to the illuminated thistles was as much as 62 percent lower, and they had less fruit. The pollinators, mainly moths, simply never reached many of the flowers, and the blossoms were literally fruitless.

The Fireworks of the Sea

On an August weekend on Sweden's west coast in the 1990s, I anchored a boat in a natural harbor in the company of two friends. On this night everything was completely still, and the late-summer darkness blurred the line between the surface of the sea and the night air, with the temperature of the water the same as that of the air. Diving from the boat into the water felt like throwing oneself into space. Stuck in my memory are the fireworks of blue light that exploded the same second my body cut through the surface. Every microscopic dinoflagellate in the water reacted instinctively by lighting its built-in lamp, and I dragged a track of light after me in the dark salty water.

We experience this phenomenon, called sea fire, in late summer and early fall, while the sea is lukewarm. The little flashes of the dinoflagellates scare off predators, such as copepods, which are small crustaceans. The light is triggered by touch, but also by fragrance. The light can also attract larger fish that eat the copepods. The dinoflagellates are not the only organisms that glow in the sea. Algae, sea squirts, crustaceans, starfish, worms, jellyfish, mollusks—there are living creatures everywhere in our oceans with the ability to produce light, if not by themselves then in cooperation with bacteria. Just among fish, we know of about fifteen hundred glowing species. As mentioned earlier, this phenomenon is called bioluminescence.

The dark and unknown deep ocean is a world completely different from our own, where light only comes for short visits. Life shows itself in flashing streaks and blinking nodes, and in between this, it is completely black. To our eyes it would be perceived as a ghostly and foreign dimension, a place not meant for us. But being able to distinguish silhouettes in the feeble daylight trickling down from the surface, or seeing other organisms' sudden fireworks, could mean the difference between life and death.

The legendary giant squid is equipped with huge eyes with a diameter of up to almost eleven inches, the largest in the entire animal kingdom. It took some time for scientists to find out what the large eyes are good for—detecting large objects, whales. The main adversaries of the giant squids are sperm whales. Reaching to sixty-five feet long, these cosmopolitan whales can dive one and a quarter miles deep in just one breath and can easily devour a squid. But every movement from the whale generates tiny flashing lights from the organisms around it. The bioluminescence from small marine animals and microscopic algae is a warning sign for the giant squids.

But even in clear water light disperses so quickly that objects, no matter how big they are, seem to disappear when viewed from a distance. Humans can only see to a distance of around thirty-three feet, even in the clearest of waters. The giant squid, however, with its plate-size eyes, can perceive the cloud of light caused by a sperm whale at up to almost four hundred feet. The light at that distance has faded out and is nothing but a cold, blurry glow. The sperm whales, in return, hunt with the help of echoes. The whales' clicking sounds are sent out at an incredible 230 decibels, in a sound wave strong enough to break the chest bones of humans and kill them. Earthquakes are the only other phenomenon that can cause sounds that powerful. The whales' sounds bounce up against potential prey, the echoes revealing what's hiding in the deep.

Ocean life is several hundred million years older than life on land and still not fully explored. The organisms in the ancient oceans probably also used bioluminescence to communicate and confound. But only in the dark abyss or in pitch-dark night is it possible to experience these fireworks. The dinoflagellates have a strict circadian rhythm, and it is only after sunset that they glitter on the surface waters. At any light disruption the effect is gone. Take a nighttime swim in the glow of streetlights and the sea-fire effect won't happen. To study these organisms, scientists need to swap day for night, either their own or the dinoflagellates', or simulate a true night.

At the Scripps Institution of Oceanography in San Diego, where the California sun blazes, there is utter darkness in the inner laboratories. Only faint red lights indicate the way to the laboratory benches, where dinoflagellates flash like sparklers in all kinds of Erlenmeyer flasks (special flasks with a flat bottom, conical body, and a short neck), jars, and glass containers. Here the dinoflagellates thrive in the safe embrace of darkness so scientists can study what happens on a chemical level when light is triggered, partly out of sheer curiosity and partly because the US navy is eager to learn how to trace unknown vessels in the oceans using natural biolight. Thanks to the light from microscopic life, satellites in orbit can detect ships and submarines as they are revealed by half-mile-long tracks, like chalk lines through the depths.

The light-emitting protein luciferin, which brings out the glow in sea fire, glowworms, and fireflies, is also used in medicine, as an indicator in the study of cancer cells. Luciferin has also been used to track bacteria in produce and to look for life in space. The name originates from Lucifer, the bearer of light, even though the name today is more associated with the dark. The moon goddess Diana is also called Diana Lucifera, thanks to her guidance during nights of the full moon. Nowadays we no longer need to collect animals with bioluminescence to extract the protein as it can be mass-produced by chemical means.

But we are a long way from thoroughly understanding the natural chemistry of light created by creatures of the world in darkness.

Where the Sea Waits

In the BBC series *Planet Earth II* there's a well-known scene in which sea-turtle eggs hatch on a beach. In the background you can see the moon reflecting on the waves, and along the horizon there's a sliver of weak light from the sun's last rays. As soon as they are out of their eggs, the tiny turtles waste no time and start crawling over the beach toward the sea and the still-light horizon in the west. They all hatch more or less at the same time, which is of critical importance to their survival. Frigate birds, gulls, crabs, and raccoons lurk everywhere, and few turtles make it into the sea. Only one in a thousand who do reach the sea will survive to adulthood, and that's the way it's been for all of time.

But then the BBC cameras pan, and on the shore behind the beach viewers can see a towering city. Streetlights, billboards, car headlights, and lights from dwellings and stores outshine the weakly glowing horizon. The turtles, which originated in the Triassic Period, more than 200 million years ago, have no reason to believe anything other than that the direction where the light is coming from strongest is west, where the sea waits. They trust their instinct to follow the light.

Some of the turtles closest to the water's edge of the beach choose the right direction and swim quickly away, happily unaware of what's going on behind them. When the camera zooms out, it becomes clear that the majority of the newly hatched turtles are moving away from the sea, toward the lights of the city instead.

Here, the film team were able to save a large number of the turtles going astray, but on other beaches around the world too many turtles are falling victim to misleading light. Ever fewer are reaching the water. On one beach in Turkey, researchers calculated that the light from nearby industries and tourist resorts meant that only two-fifths of the turtles found their way out into the waves, and in that case the beach was still relatively dark.

Nature films are often seasoned with human dramaturgy, and we project our own feelings into animal behaviors and movements as we would do with any Hollywood film. In the scene with the turtles, a single camera movement was enough to make us understand how tragically, obliviously the young turtles wandered off in the wrong direction. Suddenly we understood the seriousness of the harmful effects of light pollution on our planet. Artificial light isn't only one of mankind's most amazing inventions, but can also without a doubt be detrimental to life itself. It can mislead 200 million years of instinct in an instant.

Many different species of turtle come to the beaches of Nicaragua every year, among them the ridley sea turtle, the green sea turtle, the leatherback turtle, and the loggerhead turtle. All of them are on the International Union for Conservation of Nature Red List of Threatened Species, and they are becoming ever more rare. In Nicaragua, the situation is exacerbated by the old tradition of collecting and eating sea-turtle eggs. Less affluent people in particular, who have few other options, see turtle eggs as a chance for income. Now nonprofit forces have raised money to hire park rangers, women only, who patrol the beaches. They buy back the eggs from collectors, spread information, pick up debris, and guide the wandering young turtles at dawn. Unfortunately, the women aren't allowed to work in the middle of the night, at the time when both the young turtles and the hunters are most active. Still, they've succeeded well enough that nine out of ten eggs that would previously have ended up on a restaurant table today avoid that fate. But other dangers remain.

Those turtle hatchlings that do find their way into the water, despite everything, are born swimmers. Without parents and older relatives of their species to rely on, they must navigate out into the great ocean. They begin their journey at dusk and move out into the growing darkness. It was earlier thought that the hatchlings were left completely to the forces of the sea and weren't strong enough to swim toward goals determined by themselves, but simply drifted with the currents with chance as their only traveling companion. As with the eels' enigmatic migrations to and from the Sargasso Sea, the turtles' spectacular migrations were earlier a mystery. Unlike eels, however, turtles are relatively easy to track by satellite with GPS. It's now known that the young turtles can control their fate by swimming toward nutrient-rich areas in the world's oceans. Many turtles find their way to the same place as the eels, the Sargasso Sea, where they live under the protection of huge patches of brown algae called sargassum. In adulthood, the turtles begin the journey home to their native beaches, and with the help of a magnetic sense, like that of birds, they can with compass-like accuracy navigate thousands of miles of open sea.

Romance in the Moonlight

In a scene in the animated film *Finding Nemo* we meet sea turtles on a trip through the sea to lay eggs on beaches where they themselves were born. They get help from, among other things, the East Australian Current, which lets them surf along the coast of Australia, past the Great Barrier Reef in a southerly direction, then farther east, north of New Zealand. The stream is one hundred miles wide, moving enough water every second to fill sixteen thousand swimming pools. Many animals voluntarily or involuntarily follow the East Australian Current, one of a group of five such rotating ocean currents, of which the Gulf Stream is one.

Finding Nemo is about a clown fish who's looking for his missing son, and it was awarded an Oscar for Best Animated Film of 2003. Despite the good reviews and packed theaters, few people suspected that the film would turn the real clown fish into a billion-dollar industry. Demand for the orange-and-white fish increased throughout the world, and the number taken from the sea multiplied. The film's message—that the fish belong in the sea—partially created the opposite effect.

But at the same time, coral reefs, with their fantastic diversity, have recently been receiving at least some of the attention they deserve, and today the captive breeding of, for example, clown fish, is more profitable than the wild capture of them. The Saving Nemo Conservation Fund was founded in 2013 to protect the clown fish and contribute to the survival of coral reefs. The foundation conducts both research and education and is based at Flinders University in Adelaide, South Australia.

The clown fish is also called the anemone fish because it spends the greater part of its life among the tentacles of a sea anemone, which is something that it wouldn't occur to other animals to do. The anemone is a kind of nettle, just as corals and jellyfish are, and fires its nettle cells at anything that comes too close. The poison burns like acid and paralyzes smaller animals, which the sea anemones then eats. This poison also appears to kill cancer cells, making the anemones highly interesting for medical research. The clown fish fares well against this poison thanks to a nettle-resistant mucous layer. The mucus gets activated the first time a clown fish fry touches an anemone, and thereafter the fish can live safely within the tentacles. The species benefit from each other. The anemone protects the clown fish, and the fish shares its scraps and chases off the butterfly fish, which would otherwise eat the anemone.

Nemo, the main character in *Finding Nemo*, isn't like the other clown fish. He is drawn by the steep drop-offs out into the deep sea and adventure beyond his reef, which leads his usually-so-cautious father across the vast ocean to find his son. Nemo's real-life cousins are more like his father, never moving more than a few yards from the sea anemone. Sometimes a dozen fish live in the same anemone, but never more than one sexually mature couple at a time, a great matriarch and a dominant male who is subordinate to her. The two hold the other fish in line and seek to maintain their primary roles within the group. If the matriarch dies, her partner changes gender and becomes female. A younger fish can then become the dominant male. And when the light of the full moon breaks through the waves and finds its way through the tentacles in the anemone home, the couple plays.

The games of the clown fish are controlled by natural light and darkness, and everywhere along the reefs of the world the mating dance of the clown fish can be seen in moonlight. The eggs are placed in safe storage inside defensive walls of nettle cells and always hatch a few hours after sunset, never in daylight or even in the weak light of dusk. Darkness is absolutely crucial for the future of the clown fish.

Researchers at Flinders University and the Saving Nemo Conservation Fund have studied clown fish for several years and shown that even a small amount of unwanted light interferes with their mating cycle. If it's too bright, no fry are born. Not a single one. But as soon as it gets dark, the eggs hatch, and small, small transparent fry float to the surface. There they remain for a couple of weeks, growing until they are young clown fish, ready to swim down to the reef and look for their own sea anemones to settle in, to eventually become half of a dominant couple and play in the full moonlight. Artificial light near coral reefs threatens generations of clown fish.

All along the world's coasts, cities spread out and beaches attract tourists seeking experiences high on their bucket lists. Snorkeling through coral in a multitude of fish above anemones, algae, starfish, and crabs is a magnificent thing to be a part of.

As pleasantly vivid and colorful as the reefs are, the waters that surround them feel equally dizzying. Where the drop-offs suddenly emerge and the ocean swallows the light from above, there begins eternity. As a human being, it's easy to understand the clown fish's hesitation about the deep blue of the abyss, but also the allure that Nemo feels. What's really out there? As a tourist you want to get as close as possible, close to both the lively reef and the precipice over nothingness beyond that. Hotels are built directly alongside the water, with neon signs that beckon us to entertainment and spotlights that illuminate the promenades and luxurious bungalows placed directly on top of the reefs. Not infrequently, these luxury suites have glass floors to make the travel experience complete. The feeling of living on the reef, of looking out on the bustling life there and out toward the blue depths of eternity from the safety of your own sea anemone is enticing. It's dazzling and makes us giddy and we're happy to pay for that feeling.

But light leaks out through the glass of the floor, even if it's only faint, atmospheric lighting. From the boardwalk, spotlights sweep around, glittering off the water, and the waves reflect the facades of the always-illuminated hotels. Dusk is protracted and never leaves room for any real night. There's no darkness.

The clown fish doesn't know when it is time to play, the laid eggs remain unhatched at the bottom of the anemone nest, and the ancient interplay of the fish with the sea anemones slowly ends. We humans are a step further away from a cure to cancer when sea anemone populations decrease alongside those of the clown fish.

The Saving Nemo Conservation Fund—defender of the clown fish and coral reefs—believes that the problem is not unique to the clown fish. Many other reef fish behave in similar ways, having crucial periods of hatching during the night, which is why light pollution, along with warmer sea temperatures, is an important piece of the puzzle in understanding why so many of the earth's coral reefs are threatened with extinction. The earth's most colorful and bustling ecosystems are slowly turning into worn-out gray ruins.

Pale Coral

Corals are shelled animals called cnidarians that live completely surrounded by their protective shells, which slowly build into large reef formations. The shell harbors not only the coral animal, but also algae, which like all plants know the art of photosynthesis. The algae are what give the reef its resplendent colors, enticing both wildlife and human beings to investigate its twists and turns, cavities, and secret spots, searching for living things. Coral animals and algae have developed an intricate collaboration and are entirely dependent on one another for their well-being. When the algae die, the coral reefs' colors fade. This phenomenon has been known for a long time as something that recurs at intervals of just over twenty years, in particular during years when El Niño is strong.

El Niño is a regular weather phenomenon in the Pacific Ocean and the Indian Ocean, which every three to five years gets the trade winds to change. The warm surface water that usually follows the rotating ocean currents west along the equator instead gathers off the coast of South America. The usually cold waters are drained of fish, and rainstorms pull in over the steppe lands, as the rain forest dries out.

Under these kinds of conditions the warm water kills the algae in great numbers, draining the coral reef of color and thus causing malnutrition in the coral. As long as the coral bleaching only takes place every twentieth year, the reef recovers. The quick-growing coral need about a decade to regain their previous luster after a powerful bleaching. But the intervals between the coral bleachings have been markedly reduced, the currents aren't as stable as they were earlier, and the temperature on the earth is rising. Coral bleaching is happening as frequently as every sixth year, and the world's reefs are slowly being broken down. In 2017 nearly two-thirds of the Great Barrier Reef off the coast of Australia was affected by unusually warm water.

Ironically enough, it's also when the water is the warmest, in December, that the coral reproduce, at least in Australia. This happens once a year and, exactly as in the case of the clown fish, under the full moon. In the first sign that something is happening, tiny eggs float to the surface above the reef. The eggs have been released from the interior of the coral, and within a few minutes millions of them are floating in the water. The coral animals are hermaphrodites, that is to say, both male and female, and release their eggs and sperm at the same time. The more densely the corals release their sex cells—gametes—the greater the chance of fertilization. Individuals that start too early or get behind have less chance of propagating their genes. The performance can be compared to a snow globe, which, when you turn it over for a moment, is filled with whirling snow. The whole moment is a little surrealistic in tropical water at nearly ninety degrees Fahrenheit. When the millions of sex cells reflect the full moon's light, it can be seen for miles out into the night's darkness.

On the way to the surface, the gametes pair up and fertilize the egg, giving rise to a free-swimming larva and eventually a sessile polyp stage. It will be another year before the adult corals once again release their gametes into the moonlit water. Many are the researchers then sitting and waiting in boats above, partaking in the experience and getting insight into the life of corals.

The synchronized snowstorm of sex cells is dictated by, among other things, the moon's predictable cycle and by the rays of sunset's final phase. The inner clock of the corals is calibrated to the latter so as to time the mating as optimally as possible. Different coral species are programmed slightly differently, but within one species, the individuals are all synchronized. Or they were. All around the world's reefs the alarm bells are starting to go off at the wrong times, and the common mating night is becoming extended over several weeks. Possible explanations are the inflow of water that is too warm and the accompanying algal death, but also toxic pollution. Another reason may be that the moon is no longer as visible. The corals don't know when there's a new moon or full moon because the light from the world's big cities masks the clues in the sky. The combination of global warming and light pollution doesn't simply cause coral reefs to collapse; the tiny animals also find it difficult to rebuild their structures. They need the light so that their life partners—the algae—can photosynthesize, creating energy and nutrition. But the corals need the darkness, too, to release their sex cells in their unison dance, thereby ensuring future larvae, polyps, and reefs.

On the coral reefs, countless animals live in a kind of diversity that can only be compared with that of the rain forest. One of these animals is the palolo worm, a sea-brush worm that lives in the cavities and crevices of reef formations. Every year, when October turns into November, just after the full moon appears, the worms detach the terminal parts of their bodies, which are filled with sex cells. As with corals, this happens when the full moon is shining, and soon worm ends can be seen floating around the shallow waters in an almost infinite number. Pacific islanders have for centuries noted this event on their calendars, as the palolos are an important nutritional supplement.

Many other sea-brush worms, such as lugworms and rag worms, are also governed by the moon's cycle. The lugworm lives in tidal zones, and we often see traces of them, their waste, on our beaches, shaped into small, cylinder-patterned piles of sand. We find the rag worm on harder bottoms or among fields of eelgrass. The colorful two-inch-long worms build small, transparent tubes in which they live. In the spring or early summer, they undergo a transformation and prepare themselves for mating. Just in time for the new moon, when the water at the surface is at its darkest, they gather in large groups to perform a swirling wedding dance, pirouetting and spiraling. The behavior has been re-created in marine labs and in aquariums, but it requires a strict lunar cycle for mating to start. The more you look, the more marine organisms you'll find that rely on signals from the lunar cycle, often in combination with other clues such as sunlight levels or water temperature. Sea pens, fish, crabs, mollusks—each and every one of them depends on the moon's recurring changes to know when the next phase of life is to begin.

The Twilight Zone

I floated, hovering, with my arms outstretched. The midday sun splintered in the waves, giving the ancient limestone formations every conceivable color. Below me mingled fish, starfish, snails, and sea-brush worms; algae swayed slowly back and forth. I was traveling through Honduras visiting a small part of the world's second-largest reef, the Mesoamerican Barrier Reef system. The water was only a few yards deep above the corals, but as soon as I drew near the edge of the corals, I saw the deep sea come toward me. Every glance out past the reef made me vertiginous, as if I were overlooking the end of the world, over the edge of eternity. The light turquoise of the water quickly changed to deep blue, and all the contours grew blurry just tens of yards away. Sometimes I could make out shadows, as if shoals of big fish were approaching, or as if darkness itself was forming into life. I was reminded of the scenes from *The Big Blue*, Luc Besson's film about the free diver Jacques Mayol. He was the first to dive to a depth of one hundred meters without breathing equipment.

I saw before me how his headlamp grew all the more desolate the farther down he went, and the way his calm leg strokes took him deeper and deeper downward, unconcerned with the eternity that simply grew around him. I'm not particularly afraid of heights, nor of the dark, but this made me light-headed and nearly panic-stricken. Even so it was an incredibly fascinating feeling to gaze down at the abyss.

The light from the surface decreases rapidly in the sea; by about 220 yards deep not much of it remains. But it's still possible to measure a faint, single-hued blue-green sliver of light as far down as three-fifths of a mile. We would experience it as pitch darkness, but the zone extending down that far is nonetheless called the twilight zone of the sea. The eternal night first begins below it. At those depths, the only light to be seen comes from the organisms that create it themselves with the help of bioluminescence. The eyes that peer out into the endless darkness respond to the slightest glimmer of light, even ten times weaker than what we are able to pick up. What we consider to be total darkness actually has many nuances.

Twice a day, the earth's largest migration takes place. Plankton, crustaceans, mollusks, small fish, and a wide range of other organisms move between the dark volumes of water in the ocean depths and the lighter ones at the surface. Not only in the seas, but also in the lakes. Every night, millions of animals rise to the surface and then slowly sink back down again when dawn arrives. All these organisms have consistent internal clocks, governed by the earth's rotation and the changes in light throughout the day. Their inner rhythm moves at a predetermined pace, but every day the time is calibrated by the light, so that the movement matches up with the hours in the day.

At the poles, where the contrast between summer and winter is the greatest, the sea's rhythm pauses during the midnight sun. In the winter, when it's constantly night, the moon takes over the director's role and regulates the time with its orbit through the sky. The monthly reappearance and bright shining of the full moon also affects the oceans—wherever on earth it appears, the migration is temporarily halted. Contrastingly, a solar eclipse sets the sea in motion in the middle of the day.

Changes in natural light therefore affect entire ecosystems because darkness is synonymous with security. As daylight penetrates farther and farther into the depths, the animals sink into the protective black waters. Darkness provides a way for small, often microscopic, organisms to avoid being eaten by predatory fish. Typically, these organisms are at their most vulnerable when the moon temporarily appears, or early at dawn then they are suddenly clearly visible in the first rays of the morning sun.

We see the need for the protection of darkness further up the food chain, too. Eels only migrate when the moon is shaded and when man-made lights are out. Illuminated waterways make the fabled fish hide in the sediment on the bottom and wait out the seemingly eternal day. On a river outside Gothenburg, Sweden, where eels have been counted and their movements mapped, considerably more eels pass by when the moon is below the horizon than when it's shining. During one of the eel-counting nights, late in September in 2012, a power outage occurred at an old nineteenth-century factory. The whole area around the redbrick buildings built in the 1830s fell into darkness. The night became both longer and its darkness deeper; the song of parti-colored bats could be heard along the southern slopes of the river. On this night the activity of the eels was particularly pronounced.

Other, less dark-acclimated fish react to light, too. Perch, for example, sense small light changes, and light equivalent to one-tenth of that of the full moon affects their circadian rhythm. Maybe because the typical amount of light in lakes and seas is so small, the fish are extrasensitive. Migrating salmon are most frequently caught by seals when the waters suddenly light up, naturally or—what is becoming more common—unnaturally. Illuminated estuaries and harbor basins can benefit predators such as seals, at least temporarily. Humans have for a long time known how to use light in fishing. As mentioned earlier, fishermen have for centuries used lanterns to attract fish to their nets. In northern Norway, where trawlers hunt herring and the seines enclose tons of fish every day, humans aren't the only ones making a catch. Killer whales hear the sounds of fishing boats in the distance, and when they reach the lights from the ships, they're able to hunt without the use of sonar. By signaling with bubbles, like smoke signals, illuminated by the artificial light, the killer whales attract their relatives so that they can also take part in the buffet. The break in northern darkness gives killer whales an advantage as light changes the rules of the game and the balance between predators and prey. But it's only now that we're starting to think about the consequences of this.

In the sea off Wales, oil rigs and large ships are increasingly giving off light. Despite the knowledge that artificial light can affect various organisms, its effect on marine animals is still relatively unstudied. Two British universities launched a study on light pollution in 2013. Plastic panels equipped with LEDs were lowered a little below the surface of the Menai Strait between the island of Anglesey and the Wales mainland. Then the growth of various organisms was measured over time. Forty-seven different groups of organisms settled on the plates in intricate miniature ecosystems. Immobile animals such as corals and tubular sea-brush worms formed small landscapes, inhabited by crustaceans, larvae, and fish fry. But the more illuminated the tiles were, the lower the diversity. The light benefited only a few species of animals. Sea urchins and cnidarians chose dark plates, while Collembola and bristle worms thrived as well or better than usual on the illuminated plates, despite that the mating rituals of the bristle worms are governed by the variations of moonlight. With light equivalent to that of normal street lighting, the researchers were thus able to control the design of the small ecosystems.

On a larger scale this dynamic is concerning. About one hundred thousand merchant and transport ships sail between the earth's continents, with fifteen hundred oil rigs and at least a hundred larger wind farms out in the sea. Forty percent of the world's population lives within some sixty miles of a coastal area, and the pressure on our seas is constantly increasing. For the moment much of the great oceans still lie under a pristine and vast night sky, but you have to get farther and farther from land for starlight to appear in its full splendor and if the creatures of the sea are to live undamaged by man-made light.

Ecosystem in Flux

A lot of strange creatures live in New Zealand. The islands' far distance from other landmasses has allowed evolution to roar on freely for 125 million years, ever since the islands were no longer connected to Antarctica, Africa, and South America in the common landmass called Gondwana. The only two native mammals are bat species, which sometimes give their wings a rest and instead run on the ground eating both pollen and insects. In the absence of predators, several bird species have stopped flying and settled into niches on the ground. A number of insects have also added a behavior that we are not used to seeing in similar species in other parts of the world. One of the more spectacular of these is the weta. Wetas encompass several different species and genera of grasshoppers. They look like katydids (Tettigoniidae) with slightly extraprickly legs, but they lack wings and rarely jump around. Instead, they crawl on the ground, ideally at night. Wetas can grow spectacularly large and above all heavy. Specimens of almost four inches are not uncommon, and the record weight is put at nearly two and a half ounces. That was admittedly a female with eggs, but no less impressive. The weight matches that of a sparrow or five New Zealand bats. In Sweden we'd need to put an extra stamp on a letter of that weight to send it.

The weta's nocturnal habits reflect that in earlier times there were large birds that would more than happily eat insects of the weta's size. With few exceptions, the birds were diurnal, so wetas preferred the darkness of night to gather food. They still had enemies, including a nocturnal tuatara, a reptile unique to New Zealand that is not closely related to either contemporary lizards or crocodiles, but has been sitting all alone on its branch of the family tree for at least 225 million years. But the ground was still by far the safest place for wetas at night, at least up until today.

In the 1800s, when Europeans started coming in ever larger groups with their pets, the number of predators increased markedly. Cats, in particular, have been hard on New Zealand fauna. Few of the wild animals had the necessary defenses against these new foes, and many native species, including the large grasshoppers, have found it increasingly difficult. Their worst enemies today are rats, which seafarers—albeit highly involuntarily—brought to New Zealand as early as the 1700s. Rats, like so many other mammals, prefer to be out at dusk, so it has been suggested that domestic wildlife could be protected with the help of artificial light. Possibly both rats and cats would kill fewer New Zealand animals if it were a little brighter? The idea may sound reasonable, but the prey would also be affected. Night and day have followed each other in exactly the same way in the southern hemisphere as in the north throughout billions of years. All animals' rhythms have adjusted to that. The weta also reacts strongly to light, never leaving its hole until the sun has set. On nights with a full moon, the weta stays home. One experiment showed that most wetas—nearly 90 percent—completely eschewed looking for food in artificially lit environments. So it doesn't help that rats also avoid the light.

On islands where new predators appear, either naturally or introduced by humans, their prey usually quickly end up at a disadvantage. They lack the required defenses as they've never acquired these either during their own lifetimes or in the history of their ancestors. A long process of natural selection of behaviors and characteristics is required for predators and prey to reach a balance, which humans and their pets easily overturn.

Ecosystems and the balance between predator and prey are also upset if the environment is altered. When we illuminate our evenings and nights, we not only confuse the animals' circadian rhythms so that they no longer know when they should be hiding or out hunting. We also eliminate the possibility of camouflage and reveal the hiding places of both prey and predator. A tiger that hunts at dusk relies on its stripes to blend in with grass until it becomes invisible in the shadows, but in the increasing sky glow from the great cities of Asia, the tiger is becoming easier to detect. It's rare that such changes impact both sides equally.

Some species can be winners—such as rats. Other species cannot find food. In the case of the increasing numbers of streetlamps, in whose light opportunistic bats catch moths, there are two losers. The moths, which can avoid most chasing bats in the dark, have completely lost their defense when they're in the glow of a streetlamp. But those bat species that can typically outsmart moths—the barbastel and brown long-eared—are losing out through this change. These butterfly eaters belong to the most light-averse of the bats and don't have access to the buffet table under the lights but are instead pushed farther and farther away in the peripheral landscape.

Night Services

Bats were blamed for the COVID-19 outbreak and painted, as so many times before, as creatures of terror. But despite similarities to a virus found in horseshoe bats, it's not clear where the virus SARS-CoV-2 came from. Bats rarely infect people with viruses, although it can occur. Better we should learn from bats' unique immune system, which has been managing diseases for millions of years.

Anyone who's ever sworn at mosquitoes in the summer's darkness should welcome bats. A single bat can eat three thousand insects in one night, and a bat colony can make a big difference in the comfort felt on a terrace on quiet summer evenings. Bats are working toward their own ends, but we gratefully receive their services.

In Asia, rice is the most important food for billions of people, but crops are constantly threatened by attacks from insects and diseases. Every year more than 100 million tons of rice are destroyed because, for various reasons, they can't be eaten. Above the wet fields the bats do their best to reduce those losses, eating tons of insects that would otherwise have gone after the rice. Few pesticides are as effective, and few are equally organic. In Thailand alone the bats' services are estimated to be worth $100 million every year. Something similar applies to North America. Every night, more than 100 million bats take off from caves and bridges in the southern United States, and each can eat more than half its body weight in one day. The means five hundred tons of insects in a single night. Among their prey are owlet moths, whose larvae do great damage to corn and cotton crops. The bats' appetites save approximately $3 billion for US growers every year, money that would otherwise have been spent on pesticides.

Just like hummingbirds, bumblebees, bees, and moths, many bats seek out the flowers and fruits of plants, which make them important pollinators. More than five hundred plants around the world rely on bats for pollination. Many of these plants are both common and economically important, such as agave, balsa wood, mango, guava, and dates. Other plants rely on bats to spread their seeds when the bats release their droppings. In Thailand and Malaysia, the value of bat pollination of durian, or stink fruit, is calculated to be about $100 million annually. Durian, which can weigh up to six and a half pounds each, is considered a delicacy and is called the king of fruits, despite its strange, musty scent, which often gets it banned in hotels and on public transport. But its odor hasn't evolved for our sensibilities, but for those of the animals that spread its seeds, such as orangutans.

Out of the world's more than thirteen hundred bat species, about 70 percent are insectivores. As far back as the beginning of the twentieth century, it was realized that bats can keep insect populations in check, including that of those that spread disease, such as malaria mosquitoes. Malaria today is one of mankind's worst scourges, with over fifteen hundred deaths daily, and occasional attempts have been made to attract hungry bats to affected areas. The idea is more relevant than ever since resistant malaria has started spreading around the world. No one has yet done any comprehensive calculations of the value of maintaining bat populations, or of what would happen if the bat disappeared completely from artificially lit church towers, tourist caves, and city centers. However, more commercial actors have, in recent years, had their eyes opened to what bats eat and what comes out the other end.

Bat droppings, known as guano, have always been an effective fertilizer. In many places it's commercially mined, and even in Sweden you can find cans of guano at well-stocked garden stores. In our age of chemistry, however, we tend to forget such natural sources of phosphorus and other nutrients. But in 2014, when a couple, Melanie Drese and Michael Völker, took over an older vineyard in Germany, they set out to cultivate as organically and naturally as they could. So they did everything to attract bats, with darkness, water, insect-rich lands, and suitable habitats. A large colony of gray long-eared bats lives on the vineyard now, producing a great quantity of free manure. The farm's most popular wine is named Fledermaus, with a picture of a gray long-ear on the label. Their red Fledermaus from 2017, described as an acidic, mineral wine, with hints of strawberry, lime, and black currants, was made from grapes that had grown large from the nutrients in the bat droppings.

French winegrowers, too, understand the benefits of having bats on the farm. The Vin de Bordeaux commissioned a study in the southwest of France, in a wine region with medieval origins, south of Bordeaux. For three years, researchers collaborated with about twenty vineyards to settle bats in the district. Analyses were done of what the bats ate and whether they hunted near their settlements. Nearly all the colonies hunted over nearby vines. They also ate large quantities of winders, a species of moth whose larvae do great damage to the vines. The hope is that these results will lead to a drastic reduction in the amount of pesticides used in France and throughout Europe.

Initiatives such as these, as well as quantifying how much animals and plants can mean to humans, enable biologists to talk to economists about ecosystem services. These are the benefits carried out around the clock by the planet's various organisms. To a large extent this happens at night: pollination, pest control, decomposition, carbon dioxide storage, noise reduction, medicines—the list goes on. The worth that comes in aesthetics—the enjoyment of the scents of flowers at dusk, birdsong, the benefits of a salt bath—is harder to translate in economic terms, but many studies have shown how proximity to nature improves our sense of well-being.

PART III

Humanity and
the Cosmic Light

Three Twilights

Light hesitates, it's not instantaneous, posited Danish astronomer Ole Rømer (1644–1710) after having studied Io, one of Jupiter's moons, in 1675. Rømer was one of the first people to consider that light could have a speed. His observations became an important piece of our understanding of what light is. Not long after Rømer's observations the Dutch astronomer Christiaan Huygens (1629–95) put forward the idea that light is a wave, while his contemporary Isaac Newton (1643–1727) claimed that light consisted of particles. It would turn out that they were both right.

The person who usually gets the credit for having put Huygens's and Newton's models together is Albert Einstein (1879–1955), even if he was himself a little dubious about the result. Einstein calculated that light consists of particles, photons, but that it's also a forward-moving wave. That is, the light has two aspects, a dual nature, in which it's something physical, with a mass, and a moving field of energy. When light collides with something, it's either absorbed or reflected. When light bounces off an object and hits our eyes, we perceive the object.

Not until as late as 2015 could light be imaged in its two states, thereby confirming Einstein's calculations. But what we refer to as light in everyday speech is just a small part of something larger—electromagnetic radiation. One end of the electromagnetic spectrum includes long radio waves and microwaves, and at the other end are the short waves from X-rays and gamma radiation. The light with all its colors we find in the middle, but there are thus significantly more wavelengths than what we see. The size of the electromagnetic wave, the wavelength, determines whether we are able to perceive it. Our eyes are sensitive for light wavelengths between 380 and 800 nanometers long. The wavelengths in the upper part of that range we see as red light, in the lower as violet, and in between as the full color spectrum of the rainbow.

Many animals experience wavelengths outside our own abilities and perceive completely different nuances in the environment. The light with wavelengths long enough to be beyond our perception we call infrared, and we feel it as heat. Snakes can combine infrared vision with other sensory impressions to create an image of their surroundings and find warm-bodied prey more easily. Light with shorter wavelengths than what we can perceive we call ultraviolet, and it is used by insects and birds in a color world beyond our own.

Colors change over the day. When the sun is low, its short-wave rays are more absorbed by the atmosphere than the long-wave, red light. The landscape around an observer becomes bluer, and the sun and the sky on the horizon become redder. We experience a cycle of light that is alternately dominated by blue and red depending on whether it's morning, noon, or evening. At the same time the intensity of the light varies considerably. The sun at its zenith gives off light that is a billion times stronger than what we may experience during a cloudy night with a new moon.

As daylight decreases, shadows grow longer and colors paler. We say that darkness falls, until it has set over the landscape like a heavy blanket. The regulations of the Swedish Civil Aviation Administration state, "Darkness is the state that prevails between sunset and sunrise, when due to reduced daylight a prominent unlit object cannot clearly be distinguished at distances greater than 8 km."

The darkness of the night is contingent upon how many degrees below the horizon the sun is. But before the night with its more compact darkness comes on, the evening undergoes a metamorphosis through the three phases of twilight. A new kind of twilight occurs at every six degrees. As the upper part of the solar disk sinks to the west, the first of the three phases begins—the civil twilight. On a cloudless night, you can still read a book in the available light, as the strongest stars begin to appear as hard-to-identify points in the firmament: Vega, Capella and Arcturus. Polaris, which we often think of when we're talking about stars that are easy to see with the naked eye, is really far down the list of the brightest celestial bodies, but can steadily and reliably be seen in the north, which is why it has become an important benchmark and a prominent symbol.

How long the twilight lasts depends on where you're located. The farther north, the flatter the solar orbit is. During the spring and autumn equinoxes, civil twilight lasts about three-quarters of an hour in southern Sweden, and during the summer it's approximately an hour. North of the Dalälven River in central Sweden, civil twilight crosses midnight at some point in summer and becomes dawn, so it never really gets dark. Or as the iconic Swedish writer Harry Martinson wrote, "June night never really happens." In contrast, the nights of polar winter seem endless as the sun never reaches high enough to summon the day. And at the equator dusk falls furiously fast, regardless of the season, and each phase is over in just fifteen minutes.

When the center of the sun reaches six degrees below the horizon, civil twilight turns into nautical twilight. The brightest stars and the horizon are clear, which is a prerequisite for being able to navigate with the help of a sextant, hence the term. The sextant was invented in 1757 by the British naval officer John Campbell (1720–90) and has had a huge impact on shipping ever since. The idea is simple. By measuring the angle between different stars and the dimly lit horizon at specific times, you can determine your geographical position. Even today, in the electronic age, the method is an important backup out on the world's oceans.

Nautical twilight lasts for the time it takes the sun to drop another six degrees. So when the sun is twelve degrees below the horizon, the last of the three twilight phases begins—the astronomical. Weaker stars are now becoming clearer, though it's not yet completely dark; the direction of the sun is identifiable.

Halfway through the astronomical darkness, when the center of the sun reaches fifteen degrees below the horizon, it's sometimes referred to informally as amateur astronomical twilight. It's called that because most of the stars and heavenly phenomenon that are visible without the use of sophisticated equipment can now be seen. In writings about the scientist Tycho Brahe (1546–1601), born in what is now southern Sweden and later court astronomer in Prague, one reads that he was foremost in his discipline in skill with the naked eye, completely without optical aids. His working days began halfway in the nautical twilight and lasted until what is known as the hour of the wolf, deep in the night.

Recently a Tycho Brahe Museum opened in All Saints Church on Ven, the island off the southern coast of Sweden where Brahe spent much of his life in the service of astronomy. Visiting the museum, one can learn more about how the Renaissance man Brahe questioned the thousand-year-old laws that were said to govern the universe and how he named no less than 777 stars in the firmament, after seeing them only with his naked eye. Parts of his observatory, Stjärneborg, have been restored, but the night sky isn't the same now as at the end of the sixteenth century. The shimmer from all the city lights of the Öresund region cloud the eyes of those who today want to use their unaided eyes in search of stars. The light pollution in the sky rubs out galaxies and distant solar systems, as if we had used a dirty cloth to obscure the window facing the universe.

Dark Matter

Galaxies rotate too fast. They behave as if they consisted of something more than what is known to us. What affects the rotation of galaxies has puzzled astronomers since the phenomenon was discovered in the early 1980s.

James Peebles came to Princeton University in 1958, only a few years after the death of Albert Einstein, and has since then tried to solve some of the great mysteries of the universe. During his sixty years at the university Peebles has—following in the footsteps of his famous predecessor—studied the structure and history of the universe, from the big bang's hot, dense constitution to the protracted, increasingly cold condition that now reigns. Peebles, who today lives close to Einstein's old house, has come to shape our everyday image of the universe and has greatly contributed to our larger understanding of the slightly incomprehensible world around us.

Peebles's image of the universe can be likened to a dark scene where an ensemble is playing music. We only see a few musicians, who represent the visible matter. From what we hear, we know that more musicians must be somewhere out there in the dark, behind the scenes. We are becoming more confident that there are a lot of them, and we can even come up with a theoretical model of which instruments they're playing. To create what we're hearing requires twenty times more musicians than the ones we're seeing. Similarly, there is twenty times more of something unknown out there in the universe than we can perceive with today's methods of measuring. Some of that other stuff has come to be called dark matter—completely invisible particles. They don't emit or reflect electromagnetic radiation, which is required for our eye to be activated. Instead, we have to rely on the effect that dark matter has on other particles through its gravitational force, or, to continue the orchestral analogy, the harmony that arises from the tones between observed and unobserved musicians in the imaginary orchestra.

James Peebles's point of departure was cosmic background radiation, which has been present from the infancy of the universe, to partly confirm the big bang theory, and partly to conclude that dark matter not only exists, but that there is a lot of it. Only 5 percent of the universe consists of matter that we can see, and just over 20 percent is made up of the enigmatic dark matter, and the rest is, according to today's models, even more secretive *dark energy*, which is a key to understanding the origin of the universe. But that's outside my realm of understanding and the parameters of this book.

Today, space telescopes can image the background radiation studied by Peebles, revealing a structure, a pattern of varying radiation. And in Peebles's modern creation story, this is the design of the universe. The pattern shows how the first particles, with the help of dark matter, clumped together to become all the galaxies of the universe. Peebles received a Nobel Prize in 2019 for this breakthrough work!

We experience light as immediate, that what we see is happening now. In our sensory world, the distances are earth-size and close by. But the light is traveling, too, and it takes time to arrive. The speed of light is limited; we see the sun as it looked eight minutes ago and the North Star the way it appeared in the seventeenth century. We cannot be quite sure that something is even still there when we see it. The farther away we're looking, the further back in time we're seeing.

If we follow the trail of microwaves, that is, the background radiation of the universe, farther and farther out into space—and accordingly further and further back in time—as if it were a dwindling light in the deep sea, we'll eventually encounter an impenetrable wall. All background radiation originates from here. We've reached as close to the origin of the universe as we can get, some one hundred thousand years after the big bang, to a wall of hot plasma soup with unsorted, free particles forming a luminous fog. Here the microwaves stop guiding us in our journey through time.

But Peebles's universe goes further back than that, past the plasma soup, past the big bang, and extrapolates a creation backward into the unknown, before atoms and photons were created. Either a moment or an eternity in the dark before time begins, before God steps in with his every light and starts creation.

Religious stories of our origins as dark and chaotic may therefore contain a grain of astronomical truth. It may be that the period we're living in is just a parenthesis in the eternal life of the universe. This age of light and matter is perhaps only a fluctuation or temporary effect of the expanding space-time. We are born in the dark, we die in the dark. The light is just a momentary visit, yet all life depends on it.

The Measure of the Night Sky

In 1744, the New Year's night was lit up by one of the brightest comets in history, C/1743 X1, or the Klinkenberg-Chéseaux Comet. This could be seen in the sky during a few months around the turn of the year 1743–44, at the same time as France began its quickly abandoned attempt to invade England. The future astronomer Charles Messier (1730–1817) was then just a teenager, and the Klinkenberg-Chéseaux Comet was what ignited his interest. He was to spend his life as a comet hunter, or "comet ferret," as the French king Louis XV called him. But in practice, Charles Messier's work was largely about locating more things than just comets in the sky. He thought it would be easier to identify comets if he first sorted out everything else around them, a bit like sorting out all the green LEGO pieces and then in peace and quiet being able to find the blue pieces.

The first object that Charles Messier sorted out as a noncomet was the Crab Nebula, now known as Messier object 1, or M1. The Crab Nebula is an exploded star—a supernova remnant, first observed in 1054, that has today grown into a more than six-light-year-large cloud of gas. Before Charles Messier died at the age of eighty-six, he'd created a list of 103 objects that were not comets, but that could possibly be mistaken as such. The list of Messier objects was later expanded and now consists of 110 objects, which list is widely used by astronomers in a variety of contexts. The Messier objects are, among other things, excellent for determining the quality of the night sky, that is, the degree of darkness in a given place on the earth. The American amateur astronomer John E. Bortle made use of this in the early 2000s to establish a scale for night darkness—the Bortle scale. This is an assessment of how much the night sky is affected by light pollution. The scale has nine stages, where 1 corresponds to a natural, completely untouched sky and 9 is for the starless gray-orange sky we see above our metropolitan centers. The scale is based on how well different objects and phenomena can be observed in the sky, not least of all Messier objects, but also the faint ghost of the zodiacal light and the Milky Way.

In the best of night skies, rated as a stage 1 of 2 on the Bortle scale, one can see up to six thousand stars or other objects with the naked eye, including the Triangulum Galaxy, or Messier object 33 (M33). The Triangulum Galaxy is the most distant thing that can be seen without binoculars, and in a really dark sky it can be seen as part of the Triangle constellation. It may be one of the Milky Way's nearest neighbors, but it's still 3 million light-years away. This means that the light we see today from the Triangulum Galaxy started its journey to our eyes at the same time as the birth of our own genus, *Homo*.

On a starry sky classified as a 4 on the Bortle scale, one can still, with a trained eye, locate the M33 among about two thousand other objects, but when you reach level 6, this is impossible. Then the number of visible stars in the sky decreases to five hundred or fewer. And it's right there, at between numbers 4 and 6 on the scale, that the artificial light from our urbanized landscapes makes itself more and more felt, from rural areas to suburban skies. This is where the Milky Way's celestial bodies— for us, points—blur together more and more to eventually disappear from the sky completely by the Bortle scale's 7. At the top of the scale, only five to ten of the brightest objects in the sky are identifiable, and the artificial light is a thousand times stronger than it is in an unimpacted sky.

In February of 1701, when Sweden organized a party in honor of its victory in the Battle of Narva the year before, they literally didn't spare the gunpowder. One hundred thirty cannons, positioned strategically around Stockholm, fired a salute at the same time as the hymn "Te Deum Laudamus" echoed in the Great Church. When darkness fell, light arrangements were lit to pay homage the king and the Caroleans. One of these installations, erected in Brunkeberg Square, was an eighty-two-foot-high wooden pyramid draped in linen, designed by the court architect himself, Nicodemus Tessin the Younger. Yard-high letters praised King Charles XII and everything was illuminated from behind by more than two thousand whale-oil lamps. In a world without cinema, television, and neon signs, such displays were magnificent spectacles. Further light displays were placed around Stockholm, with texts of tribute to and images of the king. In the area of the city known as Riddarholm an amphitheater was erected, lit by a thousand lanterns. The king's light installations were a demonstration of power, and lighting is used in a similar way to this day. The previously mentioned giant spotlight, the Luxor Sky Beam in Las Vegas, is a clear example of that.

Author Paul Bogard begins an interesting trip through the Bortle scale in Las Vegas. In his book *The End of Night: Searching for Natural Darkness in an Age of Artificial Light*, he goes in search of natural darkness, beginning with the unofficial capital of neon lights. In Las Vegas, the Pleiades are the only Messier object that can be discerned, and artificial light outshines most things other than the moon and the nearest planets, Venus and Mars. With the help of astronomers, scientists, and other friends of darkness, Paul Bogard moves on from Vegas and ever-farther away from the light of the cities, in his search for darker night skies and more magnificent astronomical experiences.

Ultimately he ends up in the Eureka Valley, which is part of Death Valley National Park. It's one of the darkest places on the American continent. For Death Valley is not simply the hottest and driest region in the United States, and as such sparsely populated, but it's also the lowest, with several valleys below sea level, and mountains as tall as almost ten thousand feet all around. So despite its proximity to Las Vegas and major California cities, artificial light is effectively blocked from disturbing the darkness of the desert night. Bogard writes that the light of the Milky Way is so intense here that it creates shadows on the ground, and the distant light from Jupiter is more than enough to disturb his night vision. The concepts of light and darkness are redefined when the darkest sky, remote from civilization, is dazzling with its zodiacal lights and star fireworks, while the brightest of city skies creates dark nooks and disreputable back alleys. Our perception of what's light and dark sometimes jibes poorly with reality.

Saint Lawrence's Tears

Every year when the Nordic light once again begins to give way to darker nights, the earth crosses the orbit of the Swift-Tuttle Comet. The comet itself appears only every 133 years, but in its trail around the sun, a cloud of particles and dust follow it. These are small as a grain of sand or as large as huge rocks, which have been torn loose from the surface of the comet. When the earth crosses the comet's orbit, we see the particles as swarms of meteors, or shooting stars. At a speed of almost 125,000 miles per hour they hit the atmosphere, where they are heated by their friction with the air and made visible as thin, glowing white streaks in the night sky. On a clear, dark night in the middle of August when the meteor showers culminate, thousands of these hastily drawn little lines can be seen, soon to disappear again.

A few summers ago I wanted to experience the shooting stars in as undisturbed a way as possible. My family and I scrubbed our little rowboat as clean as we could and loaded it with blankets, comforters, and pillows. At midnight we rowed out on Lake

Tolken in southwestern Sweden, where we have our summer cottage. The cottage is an inheritance from my paternal grandfather, who had it built in the 1950s, in line with the health and wellness ideals of the time. A sparsely decorated fishing cottage was good for both mind and body, my grandfather contended, and he was probably right. The lake view, the dark evenings, and the lapping of the water do make it easy to unwind. Grandpa's armchair is still there, as are his coffee cups, the painting my dad did of an old boat, and the notebooks listing shallows, depths, and the best fishing spots. Next to the sofa is the gramophone with a hand crank, and over the woodstove still hang dried *hålkakor* (rounds of bread) that my grandma baked.

A bit out on the water we stayed the rowboat's oars and crept down among the blankets to let the boat slowly find its way through the almost-still water. With our gaze turned upward, we waited for our eyes to get used to the dark and for the sky to become filled with stars and shooting stars. The August night was chilly but the water was still warm, so mist formed and crept around the boat. We glided along, as if on a cloud, looking out into the universe. Soon we saw the first meteor, followed shortly afterward by the next, and before long it was hard to count them all. We could in principle follow the origin of all of the white lines to one area, the constellation Perseus. This meteor shower has been named the Perseids because of its visible origin. Sometimes it's also called the Laurentian Tears or Saint Lawrence's Tears, after Saint Lawrence, who died on a red-hot gridiron after having defended the sick and poor before the governor of Rome.

The Swift-Tuttle Comet, which drags in its wake Saint Lawrence's Tears, was discovered independently by Lewis Swift (1820–1913) and Horace Parnell Tuttle (1837–1923) in 1862. It is believed that a larger piece of the comet detached that year, giving rise to a large amount of debris. The comet is also on the list of possible objects that might collide with the earth in the near future, though the risk is nonetheless seen to be rather slight. Tuttle was otherwise a famous comet hunter in the spirit of Charles Messier, naming many celestial objects. Among other things, he's given his name to the Tempel-Tuttle Comet, which we can see in mid-November every year pulling the dust cloud called the Leonids after it. Every thirty-third year, the performance is extraspectacular, being most recently seen at the end of 1999, just before the turn of the millennium.

Just a few yards from the beach at the summer cottage had formerly stood a smoker. My grandfather had built it himself to smoke eel in. The lake was full of them, if one is to believe his stories from before. They were equally common everywhere else in Sweden, in streams, ditches, wells, rivers, and lakes. I have only seen an eel at the cottage once, when I scared up one from the clay bottom as I was out swimming. The homemade smoker is gone, but according to the local fishing association Lake Tolken still has some eels, which are otherwise becoming increasingly rare in our waters.

The eels' journey from inland and out from our coasts toward the Sargasso Sea begins in the autumn. They start moving under the protection of night when the moon hangs low in the sky, and with the help of some currents, they go to their mythic birthplace and the playground of all eels, far from Swedish waters. The eels never appear in the light and are generally regarded as mysterious creatures of the dark. My grandfather's fishing trips in the summer evenings were slow, exacting, approaching ceremonial. It wasn't the hunt for eels, or even better, pike perch, that was the important thing. Just as with the experience of the shooting stars, it's not the meteorites themselves that give the lasting memory, but instead the knowledge that something more out there exists to find in the dark, something to learn more about or just marvel at. I wonder if my children will be able to take their own children out on the lake.

The Only Moon?

Night starts at six degrees after the onset of astronomical dark-ness. But it's not necessarily completely dark. In the deepest hours of night, the sun still reminds us of its presence, not least through the moon's reflection of the sun's light. The light of the full moon is four hundred thousand times weaker than that of the midday sun, but quite enough to guide us through the night and effect everything from insects to people. The moon has been enormously significant throughout history—culturally, physi-cally, and ecologically.

Under a brightly shining moon, toads and frogs don't sing as loudly, and salamanders, beetles, and moths forgo their nocturnal escapades. The light of the moon is exposing, and many animals organize their lives according to its cycle. Certain bat species, especially those that hunt near trees and shrubs in clearings or close over the water, seem to possess a moon phobia, probably out of an inherited fear of birds of prey and animals lurking beneath the surface or in shrubbery. Other bats fly on as usual. Outside the mine by Taberg Mountain in southern Sweden, when the bats swarm late in a summer evening, or in the early autumn when they do their mating dance and reconnoiter winter dwellings, they do so without any concern for the phases of the moon.

Plenty of people claim to be affected by the moon. Particularly common are perceived difficulties in sleeping, but also disturbed menstruation, anxiety, and worry. Some yoga instructors advise against practicing, and even cancel their sessions, at the full moon, as they think we are overly affected by it. In the maternity ward at the Central Hospital in Karlstad, Sweden, a notice board lists recorded times for childbirth. The board also has a marker for when the full moon occurs because of the widespread myth that more children are born during nights with full moons than during nights with new moons. But the board provides a clear message: that's not the case. Children are born at any time.

In the media you can sometimes read that people are made a little crazy by the full moon and that the police receive more calls then. The concept of lunacy is very much alive, and the moon is behind many modern myths. But research and statistics speak against these beliefs. Madness, accidents, and crimes under the influence of the moon are only subjective, and the number of reported crimes isn't greater during full moons. As for the disturbed sleep, that's not about the moon at all. It's the light. A person who experiences worsened sleep when the moon is full may simply need better curtains.

Myths surrounding the moon are not a modern phenomenon and have held a central position in culture since far back in time. In 2008, the secret behind the Antikythera mechanism—a strange object found in the waters off Greece more than a hundred years earlier—was revealed. It's a two-thousand-year-old astronomical calendar, whose mechanical system could show the orbit of the sun, the risings and settings of the strongest stars, all the phases of the moon, and even solar and lunar eclipses. To know when the full moon was approaching was worthwhile in a world without lights because transport or army maneuvering could then happen at night, thanks to its light.

Today we're uncovering more and more of the secrets of the universe, and it's now been more than fifty years since humanity took its first small steps on the surface of the moon and with them the giant step into the space age. Yet at the same time more and more astronomical experiences and phenomena are being wiped out for ordinary people. In the illuminated skies of the big cities, where we can hardly see any stars anymore, the moon is still the visible moon and moves in the same path, following the same rhythm as in the generations before us. It secretly hides its backside, showing only its illuminated face and those flecks that throughout the ages have given rise to fantasies about both child abduction and divine punishment.

The Japanese author Haruki Murakami writes in the trilogy *1Q84* about an alternative world with an extra moon in the heavens. Whether this novel might have been the inspiration for a current Chinese project I will leave unsaid, but the extra moon might now become a reality. In Chengdu, in southwest China, an effort is underway to avoid once and for all cumbersome and expensive street lighting. Instead, a private space-research institute wants to send up an artificial moon. This satellite moon would be precisely programmed in its orbit to constantly follow Chengdu, with its 16 million inhabitants, and illuminate streets and squares by reflecting sunlight at night. The slightly crazy project has met with a lot of skepticism, but in 2018, the English-language *China Daily* wrote that Chinese engineers are planning three more artificial moons for orbit, expecting to save $1 billion in energy costs annually. The light from each moon is about eight times stronger than the original moon's light and is equivalent to the kind of light found a little bit into twilight. As in Murakami's novel series, this will change the experience of night and also the conditions for millions of humans, animals, and plants.

Not even on a moonless night is the sky completely without the indirect impact of the sun. As the earth orbits the sun, the ecliptic, light rays are reflected by space dust and form a faint, ghostly glow—the so-called zodiacal light. The space dust, consisting of micrometer-size particles from meteors and comets, generates a triangular shimmer against the underlying starry sky, and during a clear night with a new moon the zodiacal light constitutes about 60 percent of the night sky's light. The light is often depicted in art as a directed cone, like a spotlight on a dark theater stage. Many people out walking at night have been deceived by the light into believing that it is dawn, which is why the zodiacal light is also called false dawn. The phenomenon was scientifically described by the astronomer Giovanni Domenico Cassini in the seventeenth century, but has been known since the earliest days of mankind. In early Muslim texts, from the era of Muhammad himself, devout Muslims are warned about the wolf's tail, the false dawn, which appears as a vertical light in the night sky. During Ramadan, when meals can only be eaten between dusk and dawn, it's particularly important to keep that concept in mind. The real dawn arrives, unlike the zodiacal light, in horizontal light and not at all like a heavenly wolf's tail.

But in our modern world, where the glowing city skies eat into the night, it's not the wolf's tail that we should fear, but mankind's false dawn.

The Blue Moment

I'm sitting on the bus on the way out of Gothenburg, Sweden. Along with other bus passengers, lone truck drivers, and people driving cars, I form part of a huge winding line down the highway. As the road curves, the light shines from thousands of red lights against the dark blue sky. In the other direction, I'm met by white headlights, slightly sparser but in a constant stream. The newer the cars, the whiter and brighter the light. Farther out from the city the line slowly thins out, and soon it's just fragments of the circling worm remaining. When the bus turns off shortly thereafter, I see the valley, and we approach the hills that make up the town where I live. Hundreds of yellow pinpricks like a scattered pointillist painting create increasingly dense formations, but most can be discerned as individual units. In the seemingly opaque darkness outside the bus window, it's hard to see these small dots of light could actually pollute the night. But the number of light points grows and grows, and soon some of them can't be separated. If you look upward, you can see that the sky is dimly fluorescent. That's the clouds reflecting the light that the houses in the city center give off. I get off at school, it's Saint Lucia Day, and above me the Geminids are raging by, in one of the massive meteor storms that we can experience here on the earth. But I don't see them. The parking-lot lights shine in the haze of the sky, and only raindrops can be seen in the light. The weather and the lights are effectively blocking the night's captivating spectacle. But the Geminids tend to get bigger every year, and I hope I can be in a darker place with clearer weather next Lucia and see them.

Just before Lucia in 2016, I was invited to a small village in northern Sweden. On the way there, the December sun made a fleeting attempt to light up the ice over the Gulf of Bothnia, but soon gave up again. The journey north took two hours, and as a person from farther south in Sweden, the area seemed incredibly exotic with place names written in a Finnish dialect called Meänkieli, the crusty frozen landscape, and the black sky. Once there, I could almost feel that I was touching the arctic circle as an abundant snowfall began.

I was there to talk about the bats' nocturnal life at the community center during the two-week-long Night Festival, an annual event in the village in December, during which the slightly more than five hundred inhabitants go outside instead of trying to escape the darkness and meet up under the black sky accompanied by people from near and far, by relatives, returning residents, and friends. But also by curious visitors from the southern part Sweden, such as myself. The local restaurants come to life with candles and music, the sky feels clearer than ever before, and the sleigh tracks between the stores, the church, and the homes and parks don't even have time to be re-covered by snow in between trips because so many people are walking about.

Poets and speakers mingle with musicians, artists, and craftsmen, all of it happening in the spirit of darkness and night. In northern Sweden people don't mourn the summer. The winter may be harsh, but it is a part of the people's soul.

The phenomenon in which the sun never rises during the day is usually referred to as polar night. In Finnish it is called *kaamos*—a constant darkness, a long, uninterrupted night. It may sound gloomy and melancholy, but with the period also comes calm. There's not much to do then, or at least there hasn't been, from a historical perspective. So it's therefore been permissible to just be, to sleep, to philosophize, and to have conversations.

Kaamos is like the hibernation of the plants before they stretch out toward the light again in the spring. But the long polar night is actually not completely dark. Stars light up the sky like a blanket of distant Christmas decorations, and the northern lights, nature's own neon-and-laser play, come by now and then to visit. Green, purple, vibrant. I've experienced the northern lights once, a relatively faint occurrence, but still fully visible and surprising. I was standing watch in the night in the northernmost region of Sweden, high above the coast, looking out into the darkness. It was one of the many field exercises during my military service, and during the early-morning hours, a slow pulse of light suddenly rose over the horizon.

With alternating growing and decreasing intensity, green and pale violet at the edges, the lights rose up over me like gauzy, fluttering curtains and flashed faintly in slow motion. Maybe I

wouldn't have seen this spectacle if I hadn't already been star-
ing out into the pitch darkness below for the prior hour. My
night vision had had time to develop, rhodopsin had built its
molecular stacks and opened my retinas to the weakest of light.
It took me a moment to understand that I was standing under
the northern lights, aurora borealis. Although it was just a hint
of what the grander northern lights can be, I remember it clearly
as a powerful display.

Polar lights—northern lights in the north and southern lights
in the south—arise when the solar winds reach the earth's atmosphere at an altitude of ten miles. Electrons, charged particles,
fall in toward the earth in the oval formations at the poles, controlled by the earth's magnetic field. At the northern latitudes
we call it the auroral oval. The higher the solar activity and
the warmer the currents of electrons from the solar winds, the
farther from the poles we can see the effects. Normally we can
perceive the northern lights during dark winter nights in the
northernmost parts of Scandinavia, in northern Canada, and in
Siberia. Viking tales speak of the northern lights as reflections
off the Valkyries' armor when they lead warriors to Odin in his
preparations for Ragnarok.

In rare cases, we can see the polar lights as well in southern
Sweden, far from the cities. But it takes dense darkness for the
dancing electron winds to offer the spectacle. Under the gray-
yellow city skies we have to make do with knowing what's out
there, we never see it with our own eyes.

The polar darkness also offers other nuances. A blue color
settles like a weak filter across the vastness, especially in the

afternoon. If you look south, you'll see a faint redness from the hidden sun, while the landscape to the north is a deep blue rather than black. In the middle of the day, the blue light takes over and even the snow seems like something painted by Edvard Munch. *Kaamos* has come to be synonymous with the blue hours in the middle of the day that culminate in the blue moment, when it feels as if the entire landscape has been immersed in a swimming pool and then abruptly plucked out again. The short moment of blue is the winter's own twilight and can only be experienced at the edge of the arctic circle, particularly before the midwinter solstice has arrived. As I learned, one must wish to peer into the dark night to see the mysterious and beautiful light that it holds.

Yellow-Gray Sky

Following an earthquake in Los Angeles in 1994, the city was hit with a major power outage. As neighborhood after neighborhood went dark, the stars began to appear in increasing numbers above the rooftops. The local emergency services center was said to get calls from anxious people wondering about the strange light phenomenon in the heavens—the Milky Way, which hadn't been seen in Los Angeles for decades. The stories are a bit exaggerated, but still show how unusual and strange it is for us city dwellers today to see stars properly, and astronomers were grateful to tell the story. How many people actually called emergency services is unclear; however, the director of the Griffith Observatory answered the phone several times that evening.

The observatory is located in Griffith Park, Los Angeles, a little wilder equivalent to New York's Central Park. Here, well-attended exhibitions cover the various aspects of astronomy, and no less than 7 million people have looked through the eyepiece on the Zeiss telescope since 1935, even though it has been a long time since anyone could really see stars in Los Angeles. But despite the city lights, you can still see the surface of the moon, our neighboring planets, and the brighter objects of the heavens.

Gothenburg's observatory—Slottsskogen Observatory—has offered its panorama of the sky since 1929, and when Halley's Comet gave Sweden temporary space fever in the mid-1980s, the facility was expanded to its current size. These days the Gothenburg Astronomical Club meets from time to time in the observatory's kitchenette for a snack, while in the dark of the exhibition hall yet another ancient model of the starry sky glitters. Above all, each time the roof is opened to the side and the telescope is brought to life, it's still cause for a small moment of celebration in the observatory. Because even if the city sky isn't quite as black today as before, it's breathtaking to see the roof open up and the telescopic lens directed toward the unknown. The darkness falls slowly down from the ceiling and soon occupies the entire space where the telescope stands. The observatory becomes one with the universe.

Halley's Comet appears once in a human lifetime, every seventy-sixth year, and is next expected in 2061. The question is whether, by that time, there will be an observatory left in Gothenburg that can capture any possible new emerging interest in space. The Andromeda Galaxy and most of the stars have long since disappeared in the fiery yellow haze above the big city. Only one of five people in Europe can see the Milky Way daily, and in North America and Europe, nearly everyone, 99 percent, lives under a sky affected by light. Few people know real darkness or what a starry sky looks like. It's almost impossible to imagine the night that was commonplace for mankind only a couple of generations ago.

The painting *Starry Night* by Vincent van Gogh (1853–90) is one of history's most famous depictions of a living night sky. It dates back to a time just before the electric light took over the cityscape, to a time when the natural night with its star fireworks was public property. It's easy to interpret the painting's swirling depictions of stars in blue and yellow as chaos and madness, especially as van Gogh had been admitted to the sanatorium in Saint-Rémy-de-Provence in France. Like many artists, he was affected by the black bile, as they used to refer to it, the creative depression, melancholy that had been attributed by Aristotle to artists, philosophers, and poets. The myth of artists' inherent darkness has a long history. For the sake of his health, van Gogh chose to spend a year at a sanatorium in the south of France, where he created many of his most famous paintings, of which *Starry Night* is just one in a line of night motifs. Maybe this was a manifestation of his inner darkness, or simply how the night sky could be experienced as crackling and chaotic—before the entry of electric light.

Sweden's most famous night artist would be Eugène Jansson (1862–1915), whose paintings of Stockholm at night show the time just before the breakthrough of electricity, when only a few streets had been illuminated. In this brief moment in history the old and the new met. Jansson's works are often painted in dark blues with faint reflections in water of gas lamps, lamps that are now long absent from the streets of Stockholm.

In London, fifteen hundred gas lamps still spread their light, as a remnant from the time when it was mandatory for lanterns to be lit in the city between All Saints' Day and Candlemas. From Buckingham Palace along the Mall to Covent Garden the old lights are lit every night, for pure aesthetics and nostalgia. The oldest still-functioning lamps are from the late eighteenth century. Until 1976 the gas lamps were manually lit using a flame on the end of an eight-foot-long brass rod, but now they light automatically. It takes a lot of work to keep the lights burning; they need to be inspected and adjusted every two weeks so that their faint, homely glow can continue to tame the twilight. Although gas lamps belonged primarily to the Victorian era and were soon replaced by electric light, their legacy is nurtured in London. The royal family is particularly concerned with this heritage, which is why contemporary gas lamps have been set up around Buckingham Palace instead of electric light.

Electricity is one of man's most revolutionary discoveries and has fundamentally changed the way we live. Thomas Alva Edison (1847–1931) received a patent in 1880 for a commercial light bulb, and with that we entered a new era in human history. Let there be light, or as it said on a poster for Blanch's Café in Stockholm at the end of the nineteenth century, "Electric Lighting in the evening."

Industrial Light

My grandfather's father and my namesake, Johan Eklöf, grew up contemporaneous with the short era of gas lamps and the emergence of industrial Sweden. He was born in southern Sweden in the municipality of present-day Tidaholm, whose proximity to the Tidan River and its rushing waters made it a perfect place for industries. His ancestors had worked in the quarries in the black Cambrian shale, where I myself had been looking for fossils in the late 1990s. A flour mill had stood there since the eighteenth century and eventually a sawmill, and a dyeing and a wool factory. Johan Eklöf started working at the wool-spinning mill at only twelve years old, in 1878. His day at the mill began at six in the morning and finished at eight in the evening, except during the busiest times in the autumn, when the workday lasted until ten. In the winter, you also had to come in earlier, maybe already by five o'clock, to start firing up the heat. The factory didn't take into account light and darkness, seasons, or sleep cycles. Inside the spinning mill, it was constantly dark. Johan longed for the light and for the natural rhythm of the sun, so on Sundays he went along on arranged pleasure trips to Tidaholm and another nearby town, where there was socializing and accordion music.

Two decades later, he'd worked his way up and had been honored with the title of spinning master in the region's new scutching and weaving mill. But he had no greater privileges. Admittedly, the salary was quite good at one hundred kronor a month, but when he wanted a day off to marry my grandfather's expectant mother, Jenny, his employer said no. The proposed wedding date was the same day as the salary payments, which no one but Johan could handle, so the spinning master was therefore required to wait a week.

During Johan's early days as a spinning master he grew acquainted with electric lighting. In 1895 the first wires were pulled into the factory, which was relatively early compared to many other places in the country. Before that, both the offices and factory floors had been lit by kerosene lamps, whose dim light was made dimmer still by all the dust in the air.

The new electric light was expensive, so no light was permitted to be brighter than that of sixteen wax candles. This standard doesn't meet the current Swedish Work Environment Authority's recommendations. We would find what was then considered an illuminated industrial space to be rather dark. At the end of the 1940s, my great-grandfather was interviewed for a newspaper article about what it was like to work in the textile industry around the turn of the century. He said, among other things, "And I just wonder what the safety representative or whatever it's called would say if our factories today had that sort of lighting?"

The change had only just begun. Mankind had ever so slowly started to create a world where day and night were one and the same. In time, with improving working conditions, other kinds of businesses followed suit. Shops were open longer, the range of entertainment increased, the light from storefronts and eventually televisions spread out to the growing central towns, and like moths to a light people were drawn to the cities. The future was bright or, at least, brightness was the future. Mankind's dream of defeating darkness was about to be realized, and few people reflected that light could be harmful; on the contrary, we had everything to gain from lighting the lamp of industriousness in every home.

In preindustrial society, night was considered the evil time of the day, and the dark hid all kinds of devilry. But defenders of the darkness have always existed. Some of Europe's learned priests believed that the night was as holy as the day, for God had intended it so. The idea was that humans should simply stay at home, pray, and take care of their obligations, not be out in the shadows. As early as 1662, a London-based priest reportedly said, "We won't make day of night, nor night of day."

During the Enlightenment, the author Jean-Jacques Rousseau (1712–78) wrote that God had not given his approval for street lighting. Rousseau stood strong as a critic for the ideals of the Enlightenment and believed that mankind risked losing its soul in an overly artificial world. Astronomers in late nineteenth-century Paris and London also advocated for natural darkness when they noticed that the zodiacal light and the faint stars disappeared when the smog of the city was lit by the glow of gas lamps. They were supported by the obsessively opinionated August Strindberg (1849–1912), who was not optimistic about the introduction of the electric light. In 1884, during his time in Switzerland, he wrote the work "On the General Dissatisfaction, Its Causes and Cures." In it he goes after anyone who believes that the inventions and improvements of modern times are there for anyone other than the societal elite. Strindberg contended that electric light was just another way to get the ordinary worker to work more. Besides, it could hardly be good for the eyes. He writes, "They talk about telephones, consumer associations, higher wages, oil paintings, and easier communication, they point to banks (which didn't fail), charities (with pietism and humiliation), about electric light (which destroys the eyes and lengthens the working day—for the worker), as if not all this were mere expressions of a feverish pursuit of betterment."

Strindberg was partially correct that lighting was political. It was a way to increase efficiency, but also a demonstration of strength, in the same way as were the striking installations and the presentations of candles in the eighteenth century. The light could project an image of wealth and power. The darkness could be dispelled with the light and everyday life changed for the ordinary people. Work was to be had within industry, and around the mills the city grew. And the city never sleeps.

When the Clocks Are Off

Looking at a contemporary tropical metropolis from a distance, it's as though daylight gets sucked in and is absorbed by the city center sometime around 6:00 p.m. As darkness settles over the surrounding landscape, the city begins to radiate out in all directions, as if the sun had never set behind the horizon but was instead parked inside the center. The countless lights in the city seem to form a single shining defense against the surrounding night.

At other latitudes with clearer seasonal changes—such as in northern Sweden—you don't see the same distinct contrasts during the summer. Dusk seems to go on until dawn, and the city light is less conspicuously spread out into general dimness. As winter approaches, however, you can see the phenomenon here similar to that in the tropical cities. In the north, the snow has always reflected the northern lights, starlight, and moonlight and lightened up the short days of winter. These days, the snow reflects light from many other sources also, such as streetlamps and headlights, and a German study has measured the extent of this. Different places in the world were compared—Berlin's suburbs, a beach on the Baltic Sea in Latvia, and north of the arctic circle in Finland—and snow-covered streets were found to reflect streetlights and illuminate the sky with 33 percent more light than in areas with neither snow nor artificial lighting. If it's cloudy, the artificial light is reflected back to the ground, and within the city we experience it as if we had a double full moon every night. Downward-facing, shielded lights used to minimize light scattering work worse when snow is on the ground.

Farther south, such as in my own old hometown of Gothenburg, you could probably say that we measure snow in length rather than in depth. In winter, we move along sepia-toned streets as if in a colorized, blurred photograph. The yellow streetlights are reflected in even the smallest drop of water, creating an orange-gray mist just in time for the time of year when the Liseberg amusement park is decorated with matching pumpkins. Many people say that they get depressed in the constant darkness of the hazy west coast of Sweden in winter, while others like the opportunities the season offers to escape under a blanket without a guilty conscience. I can't help thinking about how we residents would have reacted if the city had proposed real darkness instead of all these shades of gray-yellow. Would we have felt better or worse?

On the cusp of spring in Gothenburg—and maybe everywhere else throughout Sweden—a phenomenon occurs that has caused many foreign visitors to take note. At bus stops, along south-facing walls, on streets and squares, people stop, turn their heads up toward the sky, and stand quietly for a moment. It seems almost religious, though their eyes aren't seeking God, but the oldest of gods, the sun. When the sun's rays hit our skin, the body produces vitamin D, which, among other things, helps us to make use of calcium and so strengthens our skeletons. People staying in the dark for long periods risk not getting sufficient natural vitamin D, and maybe that instinctively prompts us to turn toward the sun. Our bodies simply require the strong sunshine at a chemical level.

The bright blue and violet rays of the sun during the day don't just help the body to form vitamin D. As soon as our retinas are hit by the morning's sun, photon swarms signal the nerves to the suprachiasmatic nucleus—the node of the brain for the circadian rhythm. From here, among other things, the pineal gland, which is responsible for the body's sleep hormone, melatonin, is controlled. Melatonin is transported via blood and spinal fluid out to the cells. Daylight keeps the melatonin at a low level, and we feel active and alert. When natural outdoor light decreases and changes color, the amount of melatonin increases. Humans, just like other animals and plants, react differently to different types of light. The blue light means it's day, the red means it's evening. While it's more complicated than that, more than anything else the bluish daylight resets our internal clock, through the cryptochromes, the light-sensitive proteins, and indicates that the day can begin again.

Our built-in food and sleep clocks also follow a daily cycle of about twenty-four hours. But most of us have a cycle that runs a little longer, about fifteen minutes extra. If the inner clock is allowed to tick along freely, we'll slowly shift our day, which happens with people who've been trapped in the dark for a long time as well as people who are completely blind. So blindness can be accompanied by sleep difficulties or just difficulty living in the same rhythm with the rest of society because the body feels completely natural resting a little bit later every day. And before you know it, we're awake in the middle of the night. Some people have a somewhat shorter rhythm, and they are typically morning people.

For many animals, in addition to adapting to the time of day, it's also important to prepare for hibernation or for the arrival of spring. Melatonin also plays a major role here. Long periods of high levels of melatonin indicate that it's winter, and shorter periods of sleep hormone mean that the days have gotten longer and lighter. Then it's spring, with completely different conditions.

Between sunset and midnight the levels of sleep hormone steadily rise, which triggers a variety of reactions in the body. We get tired, the body prepares for the night's sleep cycle, the brain recovery, and the processing of the day's impressions. Body temperature drops, our metabolism decreases, and we become less hungry. The latter is because melatonin triggers another hormone, leptin, which alerts us about our energy stores and how we should manage them. During the night, leptin rises, then sinks again after sunrise. Leptin follows in the wake of the melatonin waves and regulates our appetite, rhythmically and regularly. This was particularly important in the earliest days of mankind, when we needed to conserve energy and couldn't go out looking for food in the middle of the night. Only in the morning did the body tell us that it was time to eat.

Among fifteen- to twenty-nine-year-olds, more than 80 percent have their phones with them in bed. The last thing we do in the evening is set an alarm, check social media, read emails, or just scroll. Before that we might have spent a couple of hours in front of the TV or the computer, followed by a few minutes in a bright white-tiled bathroom. Maybe we had an evening snack because the leptin, the hormone that makes us feel full, hadn't yet kicked in. When we finally turn everything off and close our eyes, we can still sense the streetlights outside the window. Passing cars may occasionally light up the room, or the neighbor's garden lighting peeks in from behind the curtain. No matter what light we expose ourselves to in the evening, we're disturbing the natural wave of melatonin, which is meant to wash over us at dusk, eventually ebbing out in time for the morning. The worst is blue light, what is most similar to daylight, but other light affects us, too.

At Harvard University it's been shown that light values of at least eight lux are sufficient to interfere with our melatonin cycle. The light value corresponds to that in the evening, the civil twilight. The disturbance means that we don't get sleepy at the right time, the brain and body don't slow down, metabolism continues as usual, and we get hungry when we shouldn't. Our quality of sleep is getting worse. Is anyone surprised?

Light Yourself Sick

That poor-quality sleep has a huge effect on us isn't news. Anyone who's taken care of small children, worked the night shift, flown across several time zones, or been out partying for a full night can attest to that. Exhaustion can be devastating, but it's easy enough to address, at least in normal cases. A good night's sleep and the body is renewed. Recurring sleep problems, however, can lead to chronic stress and depression, and even physical ailments. The body enters a vicious circle where stress and disturbed sleep go hand in hand. We become vaguely depressed. Today, tens of millions take some form of antidepressants—depression is a public-health crisis. We may not be able to cure or prevent depression all at once by cutting down on electric lighting, but we definitely increase the chances of good sleep in the long run.

To achieve this, we can resort to various tricks—some people like to sleep in an environment where natural light calibrates their interior clock; others prefer to have it as dark as possible throughout their sleep.

The important thing is for the light to vary in a cycle and that the waves of melatonin can come and go at a regular rate. Blue light during the day, red in the evening.

In addition to stress, depression, and sleep problems, obesity is also a global health problem. Obesity has many causes, but one of these is constant low leptin levels, which is a direct result of the breaking down of the melatonin circle. Simply put, we light ourselves fat. But it doesn't stop there: melatonin also controls other hormones and processes that are important for our immune system. When the Danish researcher Johnni Hansen examined seven thousand women with breast cancer twenty years ago, he drew an important conclusion. Night-shift work increases the risk of tumor formation. Hansen's work has been replicated over and over. Nurses, flight crews, and factory workers who work night shifts, who are awake at night more often than the rest of us, run a greater risk of suffering from cancer. This seems to be especially true of hormone-sensitive cancers such as breast and prostate cancer. The World Health Organization, WHO, has classified night-shift work as a cancer risk in the same category as smoking.

The causal relationships are definitely not simple, but part of the explanation lies in the night illumination. Melatonin and its effect on other hormones contributes to the inhibition of tumors, and if the biological clock is disturbed so that the melatonin wave is absent at night, its positive effects are decreased.

Researchers in Israel have seen a link between disease rates and the amount of shortwave blue light present at night. In the most light-polluted parts of the country, hormone-dependent cancers such as breast cancer were more common, while lung cancer, for example, was not. We still don't know everything about how light affects us humans and the extent to which a disturbed sleep cycle causes different kinds of health issues. But night-shift workers are clearly more at risk than other people. In our modern society, however, it is difficult to avoid night-shift work completely. Key functions depend on the availability of staff around the clock. The Central Hospital in Karlstad—the hospital that has the birth chart marked with the dates of the full moon—has invested heavily in modern lighting. The lights are governed by the time of day, mimicking natural diurnal variation. In the middle of the day, the light is white, with a large proportion of blue wavelengths. As evening progresses, blue light decreases in favor of redder shades, just as it would during a natural sunset. The intensity also varies, from a faint dawn-like light that slowly becomes more intense between morning and afternoon and then gets more and more subdued again. Night lighting in all areas of the hospital also depends on what's going on. Where there is no need for bright light, there's simply no light. The lights are set so that night staff can see to do their work, but the light lacks the blue component entirely. The daytime running lights are preprogrammed, but they can be adjusted as needed. Patients are said to sleep better, as do the staff once they get off their shifts. Karlstad's model, which is also used in other places, shows that we can regulate both the amount of light as well as its color to meet our needs for both light and darkness. Now we need to transfer these technologies into more environments where artificial lighting endangers human health.

PART IV

In Praise of Shadows

Like a Balm for the Soul

One might wonder why a book about darkness has been as much about light as this one has, and I've asked myself that question as I've worked on it. But in the texts that I've come across on the subject, darkness is only defined by light. Darkness is seen, quite simply, as an absence of light, in the same way that an absence of sound is often used to define silence. According to this kind of an approach, darkness is a kind of primordial state, ceasing to be darkness at the moment visible light is present.

I've been guilty of this myself, having used such a definition on several occasions, yet my view is that darkness has an independent worth. Not least of which, it can be as equally a concrete experience for our senses as light. It can creep over us, it can be enveloping, restful, and frightening. If we were in a room completely without light, we would probably not describe it as just dark but maybe as pitch-black or coal black. In the same way that a room lit by a candle isn't perceived as particularly bright, but rather as dark or at least semidark. Although strictly by definition, it's light that can be measured, the experience is often marked by that of the darkness. Darkness can fall; it can be intrusive and dense. The intrusion of darkness, we say, as if the proximity of darkness is an assault on our existence. Emotionally and linguistically, darkness is something substantive and concrete. We use words such as *obscure* and *murky* to describe a darkness experienced between complete light and complete dark—in a literal but also in a more philosophical or metaphorical meaning. The journalist Åke Lundqvist once wrote, "Darkness is not the absence of light. Light is diluted darkness. The speed of light is often talked about, in a kind admiration. The speed of darkness is much slower, darkness falls softly and quietly, as balm for the soul."

So, darkness is a phenomenon in its own right, I posit, yet it still seems infinitely more difficult to define than light. Especially without using light as the opposite pole. In our lived world the two seem as though forever connected, therefore a book about darkness must also inevitably be about light and not least about the interplay between light and darkness. Because without light no darkness, and without darkness no light.

In many mythologies, the relationship between light and darkness is about just this subject. The two are united but also each other's opposite poles. The night is the mother of the day, as the poet Johan Stagnelius (1793–1823) wrote, referring to the Nordic goddess of mythology Natt (Night), who was the mother of Dag (Day), Jord (Earth), Sol (Sun), and Måne (Moon). In most creation stories around the world, darkness and night are the origin and chaos, while day and light symbolize life and its origin; a sun god enters, Mother Earth takes shape. We recognize this theme from the Bible:

> In the beginning God created the heaven and the earth. And the earth was without form and void, and darkness was upon the face of the deep, and the Spirit of God moved upon the face of the waters. And God said, Let there be light: and there was light. And God saw the light, that it was good: and God divided the light from the darkness. And God called the light Day, and the darkness he called Night. And the evening and the morning were the first day.

God not only evokes the light, he is the light and all the good there is in the world, and in his dwelling—the church—the rooster reigns as a symbol of dawn, that is, the time when the night is defeated. Darkness is the absolute opposite of light and the antithesis of God: the devil, death, ignorance, and all the despair in the world. The prophet Isaiah, who was a bit of a pessimist, preached about life without God. He writes, among other things, about "trouble and darkness, dimness of anguish; and they shall be driven to darkness."

Man is said to be blind without light, literally as well as in the theological sense. Only by trusting in the light that God stands for can we get out of darkness, resist temptation, and flee evil. This view has affected how we look at darkness, and so as recently as April 2020, Pope Francis wrote that when we live in sin we are like human bats. We stay in the dark because the light reveals to us what we don't want to see, and when the eyes get used to darkness, we no longer recognize the light. The Christian light and the symbolism of darkness thus lives on to the highest degree.

In the magnificent library Biblioteca Medicea Laurenziana, in Florence, Michelangelo's bat sculptures guard its dark staircase. The lack of light in the hall was a conscious architectural choice, for only when the visitor reached the books would they, too, be rewarded with the light. Before that they were like bats, uneducated and blind to light and knowledge. Today's visitors, however, miss out on this well thought-out finesse, as the stairs in the name of security are outfitted with lights. But the bats remain there.

The library in Florence was built in the sixteenth century under charge of Pope Clement VII (1478–1534), who had been born into the influential Medici family, which dominated Florence's trade and politics for a few hundred years. Among other things, they helped to establish contact with China, whose habits, customs, and culture differed significantly from those in Europe. In China, for example, bats were regarded as anything but creepy, and not as images of the uneducable. Instead, they were symbols of happiness, wealth, and long life. If you study the statue of Lorenzo II de' Medici outside the Medici Chapel, also sculpted by Michelangelo, you can see a small box with a bat on his knee, inspired by Chinese mythology.

In traditional China, darkness and light are two balancing forces in one and the same entity. Light is born from darkness, darkness overtakes light every night. This is a milder form of opposites, where one is dependent on the other. In Nordic mythology the darkness of the underworld overtakes the Nordic autumn, but creates life in fresh soil every spring. This cycle is often attributed to the gods and recurs in religion after religion, a rhythm to which man has always subordinated himself, all the way up to modern times, until light pollution slowly but surely began to erect walls against the falling of darkness and its mandatory rest. Life's renewal every morning and every spring has lost a little of its magic.

In Praise of Shadows

In the 1930s, the Japanese author Jun'ichiro Tanizaki (1886–1965) published his book *In Praise of Shadows*. By then the neon age had begun in the world's large cities. Colorful billboards dominated the cityscape, and the zeal for illumination had taken a giant first step into the future. Tanizaki was concerned. Japan had a long cultural history with its own architectural identity, and Western influences were starting to take over more and more of the city. The urban landscapes of that time differed a lot from our modern ones. The cities were still relatively dark during the nights, but Tanizaki held that we were already chasing away the detail and the impressions of history in architecture, smudging out all nuanced experiences into one large gray mass.

The little book *In Praise of Shadows* has become a classic, especially within the architectural world, and Tanizaki was favored by Harry Martinson, a member of the Swedish Academy, for the Nobel Prize in Literature in the 1960s. Tanizaki's musings have laid the foundations for their own school of thought, which has led to Tokyo's turning off all the lights in the Roppongi Hills district. They've simply started from scratch with new light sources. Tokyo's light designers have worked with low light intensity, fewer and lower light posts, soft light on facades, and installment of artworks made out of pleasant lighting. Because the eye always adjusts to the lightest point in its visual field, we tend to experience everything else as darker. One single light or spotlight in a block causes the eye to adjust to that level of illumination. The retina's cone cells are still working as though it were day. The rod cells are deactivated and so is our night vision. The surroundings around the strong light turn black, and we experience a deeper darkness than we would have without the strong light. That means one single bright point can completely trick the eye.

In Roppongi Hills, efforts have been made to preserve the darkness as much as possible, making the district visible and safe with low-intensity and varied illumination. With no dazzling lights, experiencing the city is different, and the skyline is accentuated rather than its disappearing into a yellow haze. Light is being filtered, reflected, and dimmed. Shadows are alive and create a friendly and inviting darkness that comes to life in its entire muted color scale.

Jun'ichiro Tanizaki represents an older Eastern view of life where subtle details and shadows, such as the barely discernible texture and patina of different materials, constitute important parts of the full experience. Light and darkness don't oppose each other; the nuances unite the elements in art, architecture, and literature. A shadow falling on a tapestry highlights the craft, and golden drapes resting in semidarkness generate a different nuance from those exposed in light. Often in the East the shortcomings and the transience in natural things constitute their beauty, which can be juxtaposed to the generally more distinct striving of the West toward light, clarity, and perfection. A significant episode in Tanizaki's book is his praise of the old Japanese toilet.

> It's as if [traditional] Japanese toilets are made for giving peace of mind. They are always placed some distance away from the main building in the shade of a grove where it smells of fresh leaves and moss . . . and it is an indescribable feeling to sit curled up in the semi-daylight sunk deeply into meditation in the feeble light.

Despite that the Japanese soul lives on in both architecture and art, it is difficult to find a more technology-oriented city in the world than Tokyo. There, the Japanese countryside feels far away, and it would take many Roppongi Hills to fight the expansive light pollution.

The absolute contrast to the expansive cityscape of Tokyo would be the desolation of the deserts in the shade of the Andes on the other side of the Pacific Ocean. The Chilean Atacama Desert is considered one of the darkest places on the earth, and it's no coincidence that one of the first in a series of conferences held on darkness and the effects of artificial light took place there in 2012. Astronomers, neurobiologists, zoologists, and artists attended the symposium and the workshop called Noche Zero.

They were all there to discuss light pollution, which has since then become even more acute. The high altitude and the clear, dry air of the Atacama Desert—no one knows the last time it rained—constitute an almost optimal spot for stargazing and astronomical research. Observatories there have an open view of the cosmos and the infinity of the universe.

Beneath the Chilean starry sky, the perspective becomes different. When the eyes' rhodopsin is allowed to build its sensitive houses of cards out of molecules without any disturbing lights, the sky opens up. The night offers the entire spectrum of experience and depth of the starry sky, and nowhere else in the world do you feel so small, so insignificant, and at the same time so unique. The closest stars seem to dip toward the ground, forming a patterned backdrop behind the summits of the Andes. The most distant stars light the way toward the beginning of time, like lighthouses on a coast far away, far beyond the Pacific Ocean. Few people realize that this place is not unique. All these stars are out there no matter where on the earth we are. But we have created a blurred barrier of light, like an impenetrable dome covering our world, and only in the most remote places can we see beyond the lights. Out of all the stars we humans ought to be able to see with our naked eyes, for most of us only a half percent remain, a fragment. The rest have been obscured by artificial light, disappeared behind a smoke screen from human activity. They are there, but not there for us to see.

Diode Light

The incandescent light bulb is history, replaced by the white glow of modern LEDs. Light-emitting diodes, also known as LED lights, take advantage of how electrons are always striving for balance. Adding energy forces an electron farther out from its atomic nucleus; it soon jumps back, releasing the energy again. And this can take the form of photons, that is, light.

The LED light is energy efficient and can be operated off the grid, using only simple batteries or solar cells. The material from which the diode is made determines what color the light takes, and for a long time only red and green diodes were on the market, and eventually also yellow. But the physicists Isamu Akasaki, Hiroshi Amano, and Shuji Nakamura produced a blue diode and opened up the possibility of developing a strong white glow, a light that transforms night into day. For this innovation, they were awarded the 2014 Nobel Prize.

By combining modern diodes, light today can be controlled, programmed, and set in a way that was never close to possible with the incandescent light bulb. LEDs have revolutionized the market, and the price per lighting unit has fallen sharply, as has energy consumption per unit. The LED lamp has also made the professions of light artist and light designer attractive. A lot of training has gotten underway, and every major architectural firm now has lighting as an important part of its business. At first it was largely about the possibilities for increased power and quantity, as so often is the case with new technology. The amount of light used came into focus. But a lot of lights and bright lights do not necessarily mean that we see better. Strong lights along our sidewalks create tunnels of white light, and beyond them we see nothing. We can't see if someone is hiding beyond the light; we don't see the architecture of the city or the people moving in the evening. For security we increasingly illuminate things, but the illumination makes us dazzled and blind and—it may well be argued—less secure rather than more.

The human eye is amazing, but it's easy to underestimate its potential in low light and thus to overestimate the importance of lighting. Imagine a road at night where the reflective bits in the roadway guide you through the darkness. The car's own headlights reproduce the curves and twists of the road. Then you go through a smaller community. Streetlights take over the illumination, shining from above the road. The experience as you move past the lights is similar to that of a strobe light at a nightclub. Your eyes lose their night vision and their focus; suddenly the darkness past the buildings appears denser. You hold your breath, wait, and only when you're on the other side of the area can you breathe again. The light of the lamps fades away and the road is once again visible stretching before you into the distance, framed only by reflective white lines. The phrase *less is more* was coined by Ludwig Mies van der Rohe (1886–1969), referring to clean forms in architecture, but could just as easily refer to lighting. Well-balanced lighting makes it easier for the eye to focus in the twilight.

Today we have the knowledge and the opportunity to design lighting that is adapted for our eyes, light that can preserve the nuances of the night and create a vision that goes beyond dazzling lights to shape a harmonious evening experience in urban environments. LEDs can be directed and constrained in which surfaces are illuminated, thus reducing unwanted light scattering.

We can control and alter colors of the light to mimic the natural spectrum of the light during the day, and by controlling the intensity, shadows are featured, giving both a more natural and more pleasant impression.

But over the last decade, we've done just the opposite. We have turned on more and more lights, headlights, lamps, lighting strings, decorative lights, and facade lighting. Incandescent bulbs have been replaced by the cheaper and more energy-efficient LED lighting, but the anticipated energy savings have been eaten up by the sheer number of lights that have been installed. Nor have we taken advantage of the possibilities LED lighting offers in varied intensity; most of them shine brightly and white regardless of location and time. We are blasting ourselves with light.

Darkness Tourism

In our chaotic times, tourists are looking far and wide to find quiet areas, remote places under open skies, and pristine forests. The wilder, the better, although it's nice to also be comfortable. Around the world people are putting together these kinds of experiences in the dark: in Britain, southern Europe, in national parks in the United States, in northern Scandinavia, and along the Pacific slopes of the Andes. Astrotourism, that is, tourism linked to stargazing and the night sky, is on the rise.

One such example is the Kachi Lodge, in Bolivia. At an altitude of twelve thousand feet, an hour's flight from the city of La Paz, tourists are offered ancient celestial spectacles. Cottages shaped like futuristic globes, the way you might imagine a settlement on Mars, are located in the middle of Uyuni, the world's largest salt desert. Inside the white space-base-like domes, guests sleep as close to the cosmos as one can ever get to on earth, with an unobstructed view of the starry sky. The resort's restaurant, which is listed in the Michelin Guide, serves gourmet domestic dishes, and the barren cactus landscape all around offers the same experience of nature that once gave inspiration to the stories of the Inca people.

When Christopher Columbus sailed west in 1492, paving the way for the European invasion of Latin America, Spain's highest mountain, on Tenerife in the Canary Islands, fired off a salute. "Teide erupted," Columbus noted in his diary. Columbus devoted only those words to the earth's third-largest volcano. Today Teide has been quiet for more than a hundred years and this volcanic area has become both a World Heritage Site and a national park. If you take the cable car up alongside the mountain, you'll travel out into the universe a bit, and the experience of the Milky Way here is more intense than what most people on earth ever approach. As early as the 1960s, space observatories were opened on the volcano, and keeping pace with the light spreading from the dense stretches of hotels on the Canary Islands beaches, more and more tourists are looking up to the darkness in search for real night. Astrotourism has become a billion-dollar industry.

In the Nordic region, we can attract people with the spectacles of the midwinter night. Northern lights and stars in sparkling bands. Tourists seek their way to Iceland, northern Norway, or northern Sweden's Jukkasjärvi's ice hotel to see the sky's own fireworks and have experiences beyond the usual charter fare. In the middle of Lapland's mountain world is one of Europe's oldest national parks, Abisko. Here, too, you find a growing stream of darkness and northern-lights tourists. Visitors are guided through the night of the national park with only a dim red light, so as not to disturb their night vision. After a nocturnal ascension in a cable car, you reach the Aurora Sky Station, where the polar night seems to never end.

But even more everyday experiences are shielded from artificial light in some places. In the city of Helsingborg in southern Sweden, to preserve the beautiful sea views even in the evening, lamps on the promenades are adapted so that the view to the west is free from scattered light. A bit farther south in Lomma, it's written into the city's master plan that dark places have to be preserved.

On Møn—a Danish island southeast of the country's primary island of Zealand—best known for its spectacular cliffs, nighttime visits have become increasingly popular. The island's 460-foot-high white cliffs are impressively eye-catching in the distance and appear as though they're plunging down into the contrasting waters of the Sound, emerald green at the edges where mountains meet sea. But aside from its white cliffs, Møn is also known for its darkness. On a clear night five thousand stars can be counted over the island's cliffs, which puts Møn high on the Bortle scale, the yardstick for astronomical beauty. The corresponding number of stars in Copenhagen, just over an hour's drive away, is one hundred. So, the entire eastern part of Møn and parts of the small neighboring island of Nyord, too, have been made into a reserve—Scandinavia's first dark park, a nature area that since 2017 has been completely devoted to the unspoiled night. As if that weren't enough, Vordingborg Municipality, under whose jurisdiction the two islands fall, is also designated a Dark Sky Community. This means that the municipality undertakes to safeguard the night and that a strict lighting plan regulates how, where, and when there are to be lights. Only the absolutely most essential light is allowed.

Møn could have been Swedish land after the Treaty of Roskilde, after Charles X Gustav completed his legendary march over the Danish Belt and forced his hereditary enemy to capitulate in 1658. The Swedes' demands were high, but through the negotiations, the Danes got to keep the islands of Läsö, Anholt, and Møn. It's said this was because the Danish negotiators had placed beer glasses on the map so that the islands weren't visible. Maybe it was lucky for today's darkness enthusiasts because the Danish Environmental Protection Agency is a little more advanced than the Swedish equivalent in its view of light and darkness.

Even more at the forefront is France, which passed legislation in 2019 over how much light can be emitted into the atmosphere. In 2021 the law was fully implemented and regulates everything from brightness and color temperature to time of day and the coverings of street lighting. It remains to be seen how this will be implemented in practice and what the effects will be. But more and more countries are undertaking similar initiatives. In the Austrian capital of Vienna, they've started turning out the lights at 11:00 p.m., and in Groningen in the Netherlands industry and agriculture lights are regulated by law. Western Europe seems to have woken up in this regard, while the rest of the world is still at the starting block about the threat of light pollutants.

Møn and Nyord's status as Dark Sky Communities make them a part of an international movement. Every year, dark parks, special reserves, and municipalities are certified by the International Dark-Sky Association (IDA). Only places with extraordinary night skies are even eligible. The areas also need to be reasonably accessible and can be visited by tourists as well as researchers and amateur astronomers. A dark park is intended to be a cultural center in line with medieval cathedrals and ancient wonders. Unfortunately, few places are left that meet IDA's requirements. There are today about forty dark parks and about half as many dark communities worldwide. Only five of them are to be found in Europe, which is why Møn is such a unique place.

The darkness park on Møn, with its stars and the living nocturnal experiences, has attracted worldwide attention. During a visit to the park, you can join one of the many guided tours. On an overcast evening, the experience lies in the darkness itself, in the dense nothingness and its slowly emerging shades as your eyes get used to it. The guide presents how to best achieve a sense of security in the dark, how the untrained parts of your sensory apparatus can be activated, and how your body assimilates positively to the visual respite. If the sky is clear, the focus is on the stars. The Milky Way, with its sparkling pearls, runs the breadth of the heavens, and for those who visit the island in winter, a crackling fireworks of light from the beginning of time awaits them.

The world's darkness parks give hope. There's something uplifting in that, despite everything, parts of the night sky can be preserved in today's world of technology and illumination—if only we have the will. In Great Britain, several darkness parks hold festivals during the spring, winter, and around Halloween, which attract all kinds of people from all across the island kingdom.

The first city to receive Dark Sky City status was Flagstaff in Arizona, in 2001. The city had even then long been a pioneer in this area, when as early as 1958 it introduced the world's first lighting regulation, banning advertising spotlights. Astronomers were the driving force, but the city of Flagstaff developed its own ambition to be able to see the preserved night sky in an urban environment. A model was created limiting lighting on the basis of three criteria, similar to those in France's recently established light-pollution laws. First, all the lights face downward and are shielded by screens on top, with no light above the horizontal plane allowed. Second, the number of lights in any given area is limited, and third, light from the lamps must be warm, that is to say, the glow is yellow and red, as opposed to the cold bluish-white light that affects us the most. Flagstaff wants to be a role model and inspire everyone else to follow. If there aren't more places such as Flagstaff, France, Møn, and Nyord, we risk losing the night within a generation.

The King's Darkness

At the end of the thirteenth century, Magnus Ladulås, Sweden's then regent, acquired the area that today constitutes southern Djurgården, an area in the country's capital city, Stockholm. Since that time, the place has been under royal ownership. For a long time Djurgården was the court's private hunting grounds, and the word *djurgård* itself refers to *dieure*, the old Norse word for "deer." But by the end of the eighteenth century, at the time of King Gustav III, it was recreation and pleasure rather than hunting that took center stage. Rosendal Palace, built in the 1820s, was first of a series of magnificent buildings constructed, culminating in those for the inauguration of the Stockholm Exhibition in 1897. Since then, Djurgården and its surroundings have been dominated by museums, cultural buildings, and green parks; it's an outstanding area to visit for both residents and tourists.

The next big step was taken in 1995. Then the entire area of Ulriksdal-Haga-Djurgården-Brunnsviken was made into a *nationalstadspark*, a protected public area in the city, the first of its kind in the world. The idea was to preserve the combination of history, culture, and nature in an otherwise fast-growing metropolis. The project was unique in the world when the decision was made in 1994, but now Sweden has been overtaken by Finland, which has no fewer than ten national city parks.

At the end of the seventeenth century, between Uggleviken and what is now the Stockholm Stadium, the royal leader of the hunt, Johan Persson—more commonly known as Lill-Jan—had his hunting lodge. Djurgården was rural and wild, and being a hunter for the king meant large tracts of land to monitor. However, the salary wasn't good. So to increase his income, Lill-Jan opened a restaurant in his house. It quickly became popular and interest persisted. The restaurant remained long after his death, a bit into the nineteenth century. The name Lill-Jan is mentioned in the work of both August Strindberg and Astrid Lindgren, but not until 2009 was the forest in the area officially named after the old hunter and restaurateur. Where Lill-Jan once worked there's still relative darkness, despite all the footpaths, institutions, and the proximity to Stockholm's urban crowds. The bats still fly there today, too. We've been able to identify seven different species out of the eight that reside in the entire Djurgården.

The administrators of the area now plan to review its lighting and best adapt it to both humans and the needs of the animals. We need light, but we also need darkness. The bats show the way, and the next step will be to try out modern technology for ecological lighting. So, Djurgården and the king become pioneers in Sweden, and the National City Park adds darkness to the list of things to preserve. Maybe in Sweden, too, we are approaching the creation of dark parks similar to those on Møn and Nyord.

As part of the work at Djurgården, I got the unique opportunity to live in an old market stall (now a hostel and warehouse) in the popular open-air museum of Skansen in Stockholm and roam freely around the park at night listening to bats. I wasn't the only one who was awake. The otters, normally dozy in daytime, had some sort of convention every night that seemed to start shortly after sunset and not conclude until long after midnight. The wolves slunk around like proud silhouettes, and the mountain owl slowly turned its head in my direction every time I passed it. When the wind was blowing from the bay, Skansen was an oasis for bats. Northern bats flew silently along the footpaths and among the protruding heads of the seals, and curious glances followed the Daubenton's bats chasing after caddis flies. Around the Hällestadsstapeln bell tower soprano pipistrelles played, and by the Seglora Church a male brown long-eared circled. He'd set up in the steeple, hoping that some females would accept his offer of residence. But I saw no females. Skansen is too small an area and far too crammed into a quite illuminated urban landscape to accommodate any larger bat colonies. By all accounts, the brown long-eared male overwinters in Skansen alone, too. But maybe, with a little luck, he will find the company of conspecifics eventually.

Not even Skansen is free from lights and unnecessary spotlights, despite that the park usually closes at dusk. But the lighting is still fairly sparse. The lanterns designed in an earlier era emit a pleasant, dull light, and several walkways are completely darkened. The presence of bats tells us that wildlife can still thrive in the city where the water and parks remain unlit, and the whole of Djurgården will, one hopes, eventually come to sing in praise of the shadows and embody the Japanese philosophy of the living darkness.

Extinguished Conversation

A child who was afraid of the dark is alleged to have said, "It get lighter when someone's talking." Loneliness can be experienced more pointedly when you can't see anything, and the need for closeness and community becomes stronger. Many people feel insecure in the dark of night and want to have light around them. Our limited night vision has always made it difficult for us to discern what is out there among the shadows. Even blind people can experience fear of the dark, even though they should, in theory, do better than anyone else in total darkness. But just knowing that there's no light or that it's late may be enough for the feeling of being vulnerable to emerge.

Not only can too little light create a sense of insecurity in the darkness, but the wrong kind of lights can elicit the same effect. When we are dazzled by a light, it turns off our entire peripheral field of view, and darkness creeps in on us from all sides. A row of lanterns along a walkway ensures that we see where we are going, but outside that space lurks a dense nothingness. This is why some lighting designers prefer to talk about darkness, shadows, shades, and color scheme rather than of lighting, completely in line with the Japanese model in the book *In Praise of Shadows*. The idea is that security and darkness can be experienced together.

Some time ago I signed up for a live art installation, *Conversation*, which sought to find out what happens when we can't see the person we're talking to. You can call it performance art where the participants themselves created the content. In part it was a sociological perspective, in which you yourself are meant to converse and listen—not your body, not the gaze—a self freed from prejudices based on appearance. All clear visual expressions become unimportant when the interior self is given free rein.

I was blindfolded, led into a completely darkened room, and taken to a place at a table. Soon I gleaned that probably about ten people were in the room, maybe more. In front of me was a plate of food, a glass, and silverware. I could smell food, but couldn't see anything, nothing at all. The facilitator began by asking us to help ourselves to the food and to help each other choose and pour the drinks. Then she let us take over. None were allowed to say their name, none were allowed to reveal their profession. It was a bit tentative; someone started a conversation, which soon died out, and another person commented on the difficulty of knowing which of the two offered drinks he or she actually poured.

Soon the conversation got going, sometimes collectively around the table, sometimes between participants sitting closest to one another. Sometimes the facilitator asked us a question, but most of the time she sat quietly. I don't know if she had any kind of help that let her see us, or if she was also just part of the darkness. There was no obligation to speak, so it was difficult to know how many people were there. Maybe some people chose not to reveal their presence, but only to listen to the conversations around them.

We took turns guessing what the room around us looked like. At first we imagined a normal conference room, but gradually we ventured out under a starry sky and farther down in the deep sea. Our presentations came out slowly, but none were interrupted in their descriptions. In the absence of visual cues, we all seemed more inclined to wait and let the last of someone's words land and ebb out before we went on. The voices were distinct even though they were almost whispering, and the breathing and the small movements of the chairs were unique to each participant, though barely noticeable.

Could it be that we have forgotten quiet conversation, where silence, darkness, and slowness can have their place? Perhaps today it's almost considered a luxury? Psychologists' and therapists' reception areas are often lit, but for some patients, a dark room would probably be more beneficial. For someone who's hit a wall from stress, a lack of visual stimuli can be soothing. Some psychologists contend that darkness would enable deeper conversations, in which the therapist and the patient could speak undisturbed, completely without other impressions. It is often easier to open up and talk about yourself when you're not visible, not exposed to gazes and scrutiny in a penetrating light.

I experienced the art installation—to be able to talk like that in the dark, be quiet for a while, reflect, without any reminder of chores and tasks—as novel and valuable. Before electric lights, we spent time together in these kinds of circumstances, in the evenings and in the mornings, talking about the past day and maybe telling each other stories. You could sink into a corner of your own and just listen. *We Hear Each Other Better in the Dark* is a book of the collected letters of Franz Kafka (1883–1924), and perhaps it is so that conversations can be promoted by darkness. I found it appealing to not be seen in the *Conversation* art piece. I could sit any way I wanted, talk only if I wanted to, but still feel a sense of community with the others.

In peasant society, when natural light ruled the day, conversations in the middle of the night were part of normal social life. People went to bed early and often woke up for a while at midnight, then went back to sleep a second round until morning. Many children were conceived that way, during the quiet midnight break. Diaries and other notes tell us that a pipe was often lit, that someone might pick up a simpler bit of handiwork or fill a goblet with beer. A quiet conversation possibly took place before people fell asleep again. Some modern studies show that people who follow natural light and sleep cycles soon fall into this biphasic sleep pattern and that it can even be beneficial for health.

The restaurant Svartklubben in Stockholm has taken this concept of darkness for its interior. Strangers are placed next to one another, and unknown voices meet in the complete darkness. Hands grope for glasses and silverware, movements become slow and thoughtful, no pauses occur in conversations to pick up cell phones, which aren't allowed. The present seems unusually present. Sometimes the restaurateur, who is also a musician, picks up a guitar and plays a bit.

The musician isn't visible, but in the dense darkness, every single stanza, every single line of text and harmony, appears to create a level of comfort beyond the ordinary. Many people with ADHD have been said to find themselves at ease at Svartklubben, able to eat, relax, and listen without a flood of visual impressions and without the passing of time constantly making itself known. Perception of time changes in the dark, the clock seems to slow down, blend together with the room, and disappear. There's long been talk of light therapy for us northerners in the winter. But dark therapy is becoming a concept.

In the same area as Svartklubben the podcast *In the Dark with* . . . is recorded. The podcast isn't about darkness, but it's recorded in a room completely without light, and that is reflected in the conversations: they become deeper, calmer, less stressed, and more personal. After a period of acclimation, the guests relax and become themselves. They're not disturbed by any other impressions and can fully focus on what is being said and what they themselves are saying. The listener doesn't experience the darkness, but anyone who closes his or her eyes can share in it.

The Darkness in the Tunnel

Whether it's the lights at Djurgården, lit oil rigs at sea, illuminated Las Vegas hotels, or streetlights in a school parking lot, light pollution affects one of our most basic mechanisms: our inner clock. The circadian rhythm found in all beings on earth changes when the light is on around the clock. Most affected are the nocturnal animals, but all life at all levels is affected.

It will be extremely difficult, if not yet impossible, to stop the runaway temperatures on earth, to clean up our environment of plastics and poisons, and to prevent the spread of invasive species—plants or animals in the wrong places. It's markedly easier to dim or turn off the lights. Light pollution is the easiest of all environmental problems to solve, at least technically. The effects of turning off superfluous lighting would be immediate and wouldn't leave behind traces that need to be cleaned up. We, as private individuals, can, with little cost, reduce the amount of our light pollution. With light shades, downward-facing light sources low to the ground, and dim lighting, we can reduce the cities' total amount of light, as well as the artificial light scattered in the atmosphere. If we turn off lights in rooms we're not in, put timers on our garden lights, and motion detectors on our porch stairs, we'll get light when we need it and only when we need it.

Still, the whole thing is far from unproblematic. Light and illuminated environments mean safety for many people, and they may find it difficult to accept increased darkness. Probably few people would like to return to the world of preindustrial-age lighting. Our conception of our welfare society, our idea of security, our professional lives, our way of life, and our whole social context might need to be renegotiated. Light is also a symbol of wealth: in big cities and developed countries it shines the brightest and most. Just as when Johan Eklöf, my grandfather's father, installed electric lights in the spinning mill in the late nineteenth century, and just as when Charles XII was celebrated with thousands of whale-oil lamps, lights are a symbol of success. It's hard to refuse an African countryside electric lights when we in Sweden light up our spires and towers with ornamental lighting and drape strands of lights around our trees.

In recent years several countries and regions have initiated projects promoting darkness. In 2002 the Czech Republic and Slovenia were the first in Europe, and several other countries have followed suit or are well on their way to doing so: Italy, Spain, Croatia, Netherlands. In many places within the EU, guidelines and possible laws against excess light and for the preservation of darkness are being discussed.

The question is how much time we have to act. Many of the animals that live under the protection of darkness are on the verge of extinction and with them their invaluable services—pollination and pest hunting. And we humans have ever-worsening sleep and plants are aging prematurely when the night is absent.

Poets, philosophers, writers, and artists draw inspiration from the dark. In the absence of external images, we create our own in our interior, with the help of our imagination. In the theater, people talk about the black box, a portable stage room that is painted black where the actors can draw on their creative powers without any distracting impressions. Another type of more laid-back black box is established when we light candles in the autumn darkness or sit around the campfire in the summer evening. The Advent candleholders and the strings of light in December create moods rather than re-creating daylight. It's muted light we're seeking, the cozy lighting that promotes the relaxing darkness and intimate conversation.

I'm writing the last paragraphs of this book during Earth Hour, on a piece of paper, in the glow of candles. Earth Hour, the World Wildlife Fund's worldwide climate protest, kicks off at the end of March each year and was initially a call to make us aware of our energy waste. Now it can also be seen as a symbol of the fight against light pollution. In Berlin's city parks, the effect of Earth Hour is measurable: the city's light pollution decreases during the darkened hour, although far from everyone participates. At home, we pick out board games, put out freshly baked muffins, and make this recurring hour a family tradition. But why stop at once a year? The lights don't always have to be on; there is more to be found in the dark than we think. It's fascinating to experience how our eyes slowly acclimate and switch over to night vision, how the stars light up as the streetlights go out, and how our conversations deepen when we lean back and rest our eyes.

The night is quite simply our friend. We rest in darkness, in its stillness and subtle beauty. We draw inspiration from the night, beyond the Milky Way and the distant light. There's still life in the darkness of night, so let us take back the night, let us capture it.

Carpe noctem.

The Darkness Manifesto

Become aware of the darkness. The circadian rhythm is ancient and the basis of all life, but the darkness of the night is currently kept at bay. Help to counter that.

Protect the darkness. We are living in a world that seems to be awash in artificial light, but darkness is closer than you might think—merely a train trip or a walk or a turned-off phone away. Where do you find your darkness?

Preserve the darkness in your surroundings. Turn off the lights when you leave a room and let your garden rest in darkness at night. Observe how the darkness and the shadows' nuances appear.

Follow your inner rhythm. Let the darkness envelop you before you go to sleep, avoid blue light at night, and let the sun reset the day in the morning.

Discover nocturnal life. Take a journey beyond the city's dazzling skies and allow your eyes to adjust to the darkness. Watch the animals come out from their hiding places, their eyes glittering, and their silhouettes passing by. Smell how the plants' scents change, hear how the new sounds take over.

Seek out darkness. Observe the different phases of twilight and how the sun gives way to the moon and the stars. If you can, take yourself into the dark nights of midwinter and the mythical northern lights—a dazzling spectacle!

Learn more about the darkness and its importance for the survival of animals and plants. Let yourself be inspired by literature and art from the time before the LED light took over the night.

Talk about darkness with the people around you—the more we spread the knowledge about the benefits of darkness, the greater the chance that we can counteract problems caused by an excessively illuminated world.

Influence your environment and be a role model in the fight against light pollution. Inform your municipality of which streetlights spread unnecessary amounts of light, and how floodlighting may break environmental regulations. Participate in Earth Hour with your neighbors.

Seize darkness. Become its friend and enjoy it—it will enrich your life.

Acknowledgments

Someone once said that writing isn't an occupation, it's a form of self-harm. But it's not as solitary an occupation as legend would have it. Many people have helped me along the way, through the corridors of darkness, with their knowledge, insights, advice, and questions.

To get ideas and insight, facts, input, and viewpoints from every angle on the subjects of darkness and light, I talked to a variety of people before sitting down to write and also throughout the writing itself. Some people I met over coffee, others I just talked to on the phone, through a computer screen, or by email. Many of the conversations have become important passages in the book; others have inspired what can be read between the lines. But all of my meetings were equally important in the creation of this book; all these people contributed their thoughts on darkness and the necessity of darkness; their insights have been my companions throughout my work. And so I would like to humbly thank:

ACKNOWLEDGMENTS

Andreas Nordin (cultural scientist), Anna Bergholtz (journalist), Anette Nääs (actor), Brett Seymoure (lepidopterist), Calle Bergil (biologist), Cecilia Wide (physician, journalist, and naturalist), Charlotta Thodelius (criminologist), Freja Holmberg (museum pedagogue), Frida Rångtell (sleep researcher), Frida Sandström (art writer), Helen Arfvidsson (curator, Museum of World Culture), Henrik Aronsson (plant physiologist), Jarl Nordbladh (archaeologist), Jenny Lindström (zoophysiologist), Kajsa Sperling (lighting designer), Katja Lindblom (curator of the Slottsskogen Observatory), Magnus Gelang (curator, Gothenburg Natural History Museum), Mattias Sandberg (cultural geographer), Micael Björk (sociologist), Mikael Cremle (water conservation officer), Serena Sabatini (archaeologist), Susanna Radovic (philosopher), Taylor Stone (technology philosopher), Åsa Gunnarsdotter (psychologist).

Enormous thanks to the team who published the book originally in Sweden, Lena Forssén (publisher) and Nils Sundberg (editor), who tirelessly cut, edited, and polished the wording, as well as to those of you who read and proofread the book: Claes Bernes (astronomer), Emil V. Nilsson (botanist), Kennet Lundin (marine biologist), and Jenny Eklöf (biologist and physician).

For the English-language edition of the book, I'd like to thank my agents Paul Sebes and Rik Kleuver, who got the book into the hands of editors Stuart William at Penguin Random House in the UK and Colin Harrison and Emily Polson at Simon and Schuster (Scribner) in the United States, and to Elizabeth DeNoma, who translated the book into English.

ACKNOWLEDGMENTS

Finally, special thanks to Jens Rydell (1953–2021, zoologist and nature photographer), with whom I experienced and discussed bats and darkness since the late 1990s. I'll miss his company and counsel more than I can say. Without these adventures and conversations, this book would never have been written.

Resources for Further Reading and Listening

Bogard, P. *The End of Night: Searching for Natural Darkness in an Age of Artificial Light*. London: Fourth Estate, 2013.

Dark Sky Association. www.darksky.org.

Drake, N. "Our Nights Are Getting Brighter, and Earth Is Paying the Price." *National Geographic*, 3 April 2019.

"Good Night, Night." *Flash Forward* (podcast), 28 March 2020.

Francis-Baker, T. *Dark Skies: A Journey into the Wild Night*. London: Bloomsbury Wildlife, 2019.

Rich, C., and T. Longcore. eds. *Ecological Consequences of Artificial Night Lighting*. Washington, DC: Island Press, 2006.

Sources

ALAN—International Conference on Artificial Light at Night. Abstract booklets, 2014–18.

Albion, W., and H. E. Hanson. "The Destruction of Birds at the Lighthouses on the Coast of California." *Condor* 20, no. 1 (1918).

Anderson, G., et al. "Circadian Control Sheds Light on Fungal Bioluminescence." *Current Biology* 25 (2015).

Andersson, S., et al. "Light, Predation and the Lekking Behaviour of the Ghost Swift *Hepialus humuli* (L.) (Lepidoptera, Hepialidae)." *Proceedings of the Royal Society of London B* 265 (1998).

Angier, N. "Modern Life Suppresses an Ancient Body Rhythm." *New York Times*, 14 March 1995.

"Avsnitt 14: Mörker." *Konstform* (podcast), 30 January 2018.

Barnes, E. J. "The Early Career of George John Romanes, 1867–1878." Undergraduate thesis, Newnham College, Cambridge, 1998.

BBC. "The Making of Charles Messier's Famous Astronomy Catalogue." *Sky at Night Magazine*, August 2017.

BBC. *Planet Earth II*. 2016.

BBC Two. "The Secret Life of the Cat." *Horizon*, 2013.

Bennie, J., et al. "Cascading Effects of Artificial Light at Night: Resource-Mediated Control of Herbivores in a Grassland Ecosystem." *Philosophical Transactions of the Royal Society B* 370 no. 1667 (2015).

SOURCES

Bennie, J., et al. "Ecological Effects of Artificial Light at Night on Wild Plants." *Journal of Ecology* 104 (2016).

Bentley, M. G., et al. "Sexual Satellites, Moonlight and the Nuptial Dances of Worms: The Influence of the Moon on the Reproduction of Marine Animals." *Earth, Moon, and Planets* 85 (1999).

Bettini, A. *A Course in Classical Physics 4—Waves and Light.* Springer International Publishing, 2017.

Bible (King James version)

Bird, S., and J. Parker. "Low Levels of Light Pollution May Block the Ability of Male Glow-Worms (*Lampyris noctiluca* L.) to Locate Females." *Journal of Insect Conservation* 18, no. 4 (2014).

Björn, L. O. *Photobiology: The Science of Light and Life.* Springer Science, 2015.

"Blue Light Has a Dark Side." *Harvard Health Letter.* Cambridge, MA: Harvard Health Publishing, Harvard Medical School, 2018.

Bogard, P. *The End of Night: Searching for Natural Darkness in an Age of Artificial Light.* London: Fourth Estate, 2013.

Bortle, J. "Introducing the Bortle Dark-Sky Scale." *Sky & Telescope*, February 2001.

Broberg, G. *Nattens historia: Nordiskt mörker och ljus under tusen år.* Stockholm: Natur & Kultur, 2016.

Brouwers, L. "Animal Vision Evolved 700 Million Years Ago." *Scientific American*, November 2012.

Brüning, A., et al. "Influence of Artificially Induced Light Pollution on the Hormone System of Two Common Fish Species, Perch and Roach, in a Rural Habitat." *Conservation Physiology* 6, no. 1 (2018).

Byström Möller, L. "Luftfartsstyrelsens författningssamling." LFS 2007:33, Serie OPS, 2007.

Campion, N., and C. Impey. *Imagining Other Worlds: Explorations in Astronomy and Culture.* Bristol, UK: Sophia Centre Press, 2018.

Carrington, D. "Plummeting Insect Numbers 'Threaten Collapse of Nature.'" *Guardian*, 10 February 2019.

Carson, R. *Silent Spring.* New York: Houghton Mifflin, 1962.

Castellani, C., et al. "Exceptionally Well-Preserved Isolated Eyes from Cambrian 'Orsten' Fossil Assemblages of Sweden." *Palaeontology* 55, no. 3 (2012).

Chepesiuk, R. "Missing the Dark—Health Effects of Light Pollution." *Environmental Health Perspectives* 117, no. 1 (2009).

Ciach, M., and A. Fröhlich. "Ungulates in the City: Light Pollution and Open Habitats Predict the Probability of Roe Deer Occurring in an Urban Environment." *Urban Ecosystems* 22 (2019).

Cuff, M. "Lights Out? Habitat Loss and Light Pollution Pose Grave Threat to UK Glow Worms." *iNewsletter*, 4 February 2020.

Danielsson, U. *Mörkret vid tidens ände: En bok om universums mörka sida.* Stockholm: Fri Tanke Förlag, 2015.

Dark Sky Association. www.darksky.org.

Darwin, C. *The Expression of the Emotions in Man and Animals.* Fontana Press, 1872.

Dauchy, Robert R. T., et al. "Circadian and Melatonin Disruption by Exposure to Light at Night Drives Intrinsic Resistance to Tamoxifen Therapy in Breast Cancer." *Cancer Research* 74, no. 15 (2014).

Depledge, M., et al. "Light Pollution in the Sea." *Marine Pollution Bulletin* 60 (2010).

Desouhant, E., et al. "Mechanistic, Ecological, and Evolutionary Consequences of Artificial Light at Night for Insects: Review and Prospective." *Entomologia Experimentalis et Applicata* 167 (2018).

Di Domenico, A. "European Union Adopts New Guidance to Reduce Light Pollution." *Environmental Protection*, 6 December 2019.

Dimovski, A. M., and K. A. Robert. "Artificial Light Pollution: Shifting Spectral Wavelengths to Mitigate Physiological and Health Consequences in a Nocturnal Marsupial Mammal." *Journal of Experimental Zoology Part A: Ecological and Integrative Physiology* 329, no. 8–9 (2018).

Doctor, R. M., et al. *The Encyclopedia of Phobias, Fears, and Anxieties.* New York: Facts on File, 2008.

Dokken, P. "Varför är mörkret viktigt, Kajsa Sperling?" *Göteborgs-Posten*, 23 March 2019.

Dominoni, D. M., et al. "Clocks for the City: Circadian Differences between Forest and City Songbirds." *Proceedings of the Royal Society of London B* 280 (2013).

Drake, N. "Our Nights Are Getting Brighter, and Earth Is Paying the Price." *National Geographic*, 3 April 2019.

Duarte, C., et al. "Artificial Light Pollution at Night (ALAN) Disrupts the Distribution and Circadian Rhythm of a Sandy Beach Isopod." *Environmental Pollution* 248 (2019).

Edqvist, B., and J. Eklöf. *Fladdermusen—i en mytisk värld.* Bjärnum, Sweden: Bokpro, 2018.

Ekirch, A. R. *At Day's Close: Night in Times Past.* New York: W. W. Norton, 2005.

Eklöf, J. *Djurens evolution.* Stockholm: Caracal Publishing, 2008.

Eklöf, J., and J. Rydell. *Fladdermöss—i en värld av ekon.* Hirschfeld Förlag, 2015.

———. "Det dödliga ljuset." *Forskning & Framsteg*, 27 September 2018.

Eliasson, C. "Växter reagerar på ljus inom bråkdelen av en sekund." Press release. Göteborgs universitet, 31 March 2020.

Elgert, C., et al. "Reproduction under Light Pollution: Maladaptive Response to Spatial Variation in Artificial Light in a Glow-Worm." *Proceedings of the Royal Society of London B* 287 (2020).

Emlen, S. T. "The Stellar-Orientation System of a Migratory Bird." *Scientific American* 233, no. 2 (1975).

Englund, P. *Förflutenhetens landskap.* Stockholm: Atlantis, 1991.

Entomologischer Verein Krefeld. www.entomologica.org.

European Festival of the Night. www.nightfestival.se.

Falchi, F., et al. "Light Pollution in USA and Europe: The Good, the Bad and the Ugly." *Journal of Environmental Management* 248 (2019).

Falchi, F., et al. "The New World Atlas of Artificial Night Sky Brightness." *Science Advances*, 10 June 2016.

Farnworth, B., et al. "Photons and Foraging: Artificial Light at Night Generates Avoidance Behaviour in Male, but Not Female, New Zealand Weta." *Environmental Pollution* 236 (2018).

Farrington, D. P., and B. C. Welsh. *Förbättrad utomhusbelysning och brottsprevention. En systematisk forskningsgenomgång.* Brå-rapport 2007:28.

Fimmerstad, L. *Elljuset tränger undan gaslyktorna i Stockholm.* No date. stockholmshistoria.com.

Firebaugh, A., and K. J. Haynes. "Light Pollution May Create Demographic Traps for Nocturnal Insects." *Basic and Applied Ecology* 34 (2019).

Fobert, E. K., et al. "Artificial Light at Night Causes Reproductive Failure in Clownfish." *Biology Letters* 15, no. 7 (2019).

"Folkminnesuppteckning, Broby, Lund 1874." DAL 32 (1874).

Foster, J. J., et al. "Orienting to Polarized Light at Night—Matching Lunar Skylight to Performance in a Nocturnal Beetle." *Journal of Experimental Biology* 222 (2019).

Fox, D. "What Sparked the Cambrian Explosion?" *Nature* 530, no. 18 (2016).

Francis-Baker, T. *Dark Skies: A Journey into the Wild Night.* London: Bloomsbury Wildlife, 2019.

Fredelius, A. "Rätt ljus ger piggare läkare." *Ny Teknik,* 20 June 2016.

Gallaway, T., et al. "The Economics of Global Light Pollution." *Ecological Economics* 69, no. 3 (2010).

Garcia-Saenz, A., et al. "Evaluating the Association between Artificial Light-at-Night Exposure and Breast and Prostate Cancer Risk in Spain (MCC-Spain Study)." *Environmental Health Perspectives* 126, no. 4 (2018).

Gardner, J. "Fladdermöss hjälper ekologisk vinodling." *Vinjournalen,* 27 August 2018.

Garnert, J. "August Strindberg's ljus." *Ljuskultur* 4 (2012).

———. *Ut ur mörkret: Ljusets och belysningens kulturhistoria.* Lund, Sweden: Historiska Media, 2016.

Gaston, K. J. "Reducing the Ecological Consequences of Nighttime Light Pollution: Options and Developments." *Journal of Applied Ecology* 49 (2012).

Gaukel Andrews, C. "The Largest Migration on Earth Is Vertical." 2018. NatHab.com.

Gauthreaux, S., and C. G. Belser. "Effects of Artificial Night Lighting on Migrating Birds." In *Ecological Consequences of Artificial Night Lighting,* edited by C. Rich and T. Longcore. Washington, DC: Island Press, 2006.

Geffen, K., et al. "Artificial Night Lighting Disrupts Sex Pheromone in a Noctuid Moth." *Ecological Entomology* 40 (2015).

Gibbens, S. "As the Arctic Warms, Light Pollution May Pose a New Threat to Marine Life." *National Geographic,* 5 March 2020.

———. "See 'Underwater Snowstorm' of Coral Reproducing." *National Geographic,* 5 January 2018.

"Good Night, Night." *Flash Forward* (podcast), 28 March 2020.

Grenis, K., and S. M. Murphy. "Direct and Indirect Effects of Light Pollution on the Performance of an Herbivorous Insect." *Insect Science* 26, no. 4 (2018).

SOURCES

Griffith Observatory. www.griffithobservatory.org.

Grønne, J. "Tusentals meteorer målar himlen." *Illustrerad Vetenskap*, 12 August 2019.

Guilford, T. "Light Pollution Causes Object Collisions during Local Nocturnal Manoeuvring Flight by Adult Manx Shearwaters *Puffinus puffinus*." *Seabird* 31 (2019).

Gustafsson, B. "Fjärilar insamlade i ljusfälla på Naturhistoriska riksmuseets tak." No date. *Hagabladet*.

Hadenius, P. *Paus: Konsten att göra något annat.* Stockholm: Natur & Kultur, 2019.

Hadhazy, A. "Fact or Fiction: The Days (and Nights) Are Getting Longer." *Scientific American*, 14 June 2010.

Haim, A., and A. E. Zubidat. "Artificial Light at Night: Melatonin as a Mediator between the Environment and Epigenome." *Philosophical Transactions of the Royal Society of London, Series B, Biological Sciences* 370, no. 1667 (2015).

Hallmann, C. A., et al. "More Than 75 Percent Decline over 27 Years in Total Flying Insect Biomass in Protected Areas." *PLoS ONE* 12, no. 10 (2017).

Hallemar, D. "Disciplinerande ljus och förlåtande mörker." *OBS*, Sveriges Radio P1, 20 February 2017.

Hansen, J. "Night Shift Work and Risk of Breast Cancer." *Current Environmental Health Report* 4, no. 3 (2017).

Härdig, A. "Myter om månen." *Populär Astronomi* 1 (2019).

Hart, A. "The Rise of Astrotourism: Why Your Next Adventure Should Include Star-Gazing." *Telegraph*, 11 July 2018.

Heinrich, B. *The Homing Instinct: Meaning & Mystery in Animal Migration.* Boston: Houghton Mifflin Harcourt, 2014.

Heintzenberg, F. *Nordiska nätter: Djurliv mellan skymning och gryning.* Lund, Sweden: Bio & Fokus Förlag, 2013.

Hitta Nemo. Pixar, 2003.

Hölker, F., et al. "The Dark Side of Light: A Transdisciplinary Research Agenda for Light Pollution Policy." *Ecology and Society* 15, no. 4 (2010).

Hölker, F., et al. "Light Pollution as a Biodiversity Threat." *Trends in Ecology & Evolution* 25, no. 12 (2010).

SOURCES

Howard, J. "These Fish Eggs Aren't Hatching. The Culprit? Light Pollution." *National Geographic*, 9 July 2019.

Hughes, H. C. *Sensory Exotica: A World beyond Human Experience*. Cambridge, MA: Bradford Books, 2001.

Hunt, R. *The Poetry of Science: Or, Studies of the Physical Phenomena of Nature*. Reeve, Benham, and Reeve, 1849.

I mörkret med . . . (podcast). www.imorkretmed.se.

International Dark Sky Association. *Fighting Light Pollution: Smart Lighting Solutions for Individuals and Communities*. Mechanicsburg, PA: Stackpole Books, 2012.

Irenius, L. "Gläds åt mörkret—det är hotat." *Svenska Dagbladet*, 19 November 2019.

Jabr, F. "How Moonlight Sets Nature's Rhythms." *Smithsonian Magazine*, 21 June 2017.

Jägerbrand, A. K. *LED-belysningens effekter på djur och natur med rekommendationer: Fokus på nordiska förhållanden och känsliga arter och grupper*. Linköping, Sweden: Calluna AB, 2018.

Jechow, A. "Observing the Impact of WWF Earth Hour on Urban Light Pollution: A Case Study in Berlin 2018 Using Differential Photometry." *Sustainability* 11, no. 3 (2019).

Jechow, A., and F. Hölker. "Snowglow—the Amplification of Skyglow by Snow and Clouds Can Exceed Full Moon Illuminance in Suburban Areas." *Journal of Imaging* 5, no 8 (2019): 69.

Johannisson, K. *Melankoliska rum*. Stockholm: Albert Bonniers Förlag, 2009.

Josefsson, L. "Elegi över ett spinneri." *Göteborgs-Posten*, 10 October 2019.

Kachi Lodge, Bolivia. 2019. www.kachilodge.com.

Kafka, F. *Man hör varandra bättre i mörker: Brev 1918–juni 1920*. Lund, Sweden: Bakhåll, 2014.

Karlsson, B-L., et al. "No Lunar Phobia in Swarming Insectivorous Bats (Family Vespertilionidae)." *Journal of Zoology* 256, no. 4 (2002).

Kay, J. "Nighttime Lights Reset Birds' Internal Clocks, Threatening Dawn's Chorus." *National Geographic*, 6 September 2014.

Klarsfeld, A. "At the Dawn of Chronobiology." ESPCI ParisTech Neurobiology Laboratory, 2013.

Knop, E. "Artificial Light at Night as a New Threat to Pollination." *Nature* 548, no. 7666 (2017).

Konstverket SAMTAL. Jönköpings Läns Museum.

Kronberg, K., et al. *Minnet av Narva: Om troféer, propaganda och historiebruk.* Lund, Sweden: Nordic Academic Press, 2018.

Krönström, J. "Control of Bioluminescence: Operating the Light Switch in Photophores from Marine Animals." PhD diss., Zoologiska institutionen, Göteborgs universitet, 2009.

Kunz, T., et al. "Ecosystem Services Provided by Bats." *Annals of the New York Academy of Sciences* 1223 (2011).

Land, M. F., and D-E. Nilsson. *Animal Eyes.* Oxford Animal Biology Series. Oxford: Oxford University Press, 2002.

Last, K. S., et al. "Moonlight Drives Ocean-Scale Mass Vertical Migration of Zooplankton during the Arctic Winter." *Current Biology* 26 (2016).

Lawal, S. "Fireflies Have a Mating Problem: The Lights Are Always On." *New York Times*, 3 February 2020.

Leanderson, P. "Ljusföroreningar och mörkret som försvann." *Arbets-och miljömedicinbloggen* (blog). AMM Östergötland, 2018. http://arbetsoch miljomedicin.se/ljusfororeningar-och-morkret-som-forsvann/.

Light Pollution Map. www.lightpollutionmap.info.

Liljemalm, A. "Utrotningshotat nattmörker." *Forskning & Framsteg*, 15 March 2016.

Longcore, T., and C. Rich. "Ecological Light Pollution." *Frontiers in Ecology and the Environment* 2, no. 4 (2004).

Lövemyr, A. "Att skapa plats för mörker och natthimlen: Belysning och människan i stadens nattlandskap." SLU, Fakulteten för landskapsarkitektur, trädgårds-och växtproduktionsvetenskap, 2018.

Lundqvist, Å. "Krönika." *Dagens Nyheter*, 21 December 1990.

Macgregor, C. J., et al. "Effects of Street Lighting Technologies on the Success and Quality of Pollination in a Nocturnally Pollinated Plant." *Ecosphere* 10, no. 1 (2019).

Macgregor, C. J., et al. "Moth Biomass Increases and Decreases over 50 Years in Britain." *Nature Ecology and Evolution* 3 (2019).

Macgregor, C. J., et al. "Pollination by Nocturnal Lepidoptera, and the Effects of Light Pollution: A Review." *Ecological Entomology* 40 (2015).

Manríquez, P. H., et al. "Artificial Light Pollution Influences Behavioral and Physiological Traits in a Keystone Predator Species, *Concholepas concholepas.*" *Science of the Total Environment* 661 (2019).

SOURCES

Marsh, G. P. *Man and Nature: Or, Physical Geography as Modified by Human Action*. New York: Charles Scribner, 1864.

Mårtenson, J., and R. Turander. *Kungliga Djurgården*. Stockholm: Wahlström & Widstrand, 2007.

Martini, S., and S. Haddock. "Quantification of Bioluminescence from the Surface to the Deep Sea Demonstrates Its Predominance as an Ecological Trait." *Scientific Reports* 7 (2017).

Martinson, H. *Cikada*. Stockholm: Albert Bonniers Förlag, 1953.

McCarthy, D. D., and K. P. Seidelmann. *Time: From Earth Rotation to Atomic Physics*. Cambridge: Cambridge University Press, 2018.

McConnell, A. "Effect of Artificial Light on Marine Invertebrate and Fish Abundance in an Area of Salmon Farming." *Marine Ecology Progress Series* 419 (2010).

McGrane, S. "The German Amateurs Who Discovered Insect Armageddon." *New York Times*, 4 December 2017.

McMenamin, M. A. S. *The Garden of Ediacara*. New York: Columbia University Press, 1998.

Meravi, N., and S. Kumar. "Effect Street Light Pollution on the Photosynthetic Efficiency of Different Plants." *Biological Rhythm Research* 51 (2020).

Merritt, D. J., and A. Clarke. "The Impact of Cave Lighting on the Bioluminescent Display of the Tasmanian Glow-Worm *Arachnocampa tasmaniensis*." *Journal of Insect Conservation* 17, no. 1 (2012).

Milosevic, I., and E. McCabe. *The Psychology of Irrational Fear*. Boston: Greenwood, 2015.

Moore, M. V. "Urban Light Pollution Alters the Diel Vertical Migration of *Daphnia*." *SIL Proceedings* 27 (2000).

"Mörkertrilogin," installments 7–9. *Professor Magenta* (podcast).

Morris, H. "The Casino Light Beam That's So Bright It Has Its Own Ecosystem (and Pilots Use It to Navigate)." *Telegraph*, 24 August 2017.

Murakami, H. *1Q84*. 3 vols. Stockholm: Norstedts, 2011.

Musila, S., et al. "No Lunar Phobia in Insectivorous Bats in Kenya." *Mammalian Biology* 95 (2019).

Nagel, T. "What Is It Like to Be a Bat?" *Philosophical Review* 83, no. 4 (1974).

Nichols, C. A., and K. Alexander. "Creeping in the Night: What Might Ecologists Be Missing?" *PLoS ONE* 13, no. 6 (2018).

Night on Earth. Netflix, 2020.

Nilsson, D-E. "Havsmonstrets vakande öga." *Forskning & Framsteg*, 7 August 2012.

Nobelförsamlingens pressmeddelande för Nobelpriset i fysik, 2014.

Nobelförsamlingens pressmeddelande för Nobelpriset i fysik, 2019.

Nobelförsamlingens pressmeddelande för Nobelpriset i fysiologi eller medicin, 2017.

Nobel Prize Nomination Database. www.nobelprize.org/nomination /redirector/?redir=archive.

Noche Zero. https://lightcollective.net/light/ing/noche_zero.

Nordstrand, M. "Stökigt med fullmåne." *Örnsköldsviks Allehanda*, 7 August 2009.

Nygren, A., et al. *Ringmaskar: Havsborstmaskar: Annelida: Polychaeta: Aciculata.* Nationalnyckeln till Sveriges flora och fauna. Uppsala, Sweden: ArtDatabanken, SLU, 2017.

Nyström, J. "Koralldöden har blivit fem gånger värre." *Forskning & Framsteg*, 4 January 2018.

"Ovanligt mörker över Stockholm oroade många." *Dagens Nyheter*, 17 October 2017.

Owens, A. C. S., et al. "Light Pollution Is a Driver of Insect Declines." *Biological Conservation* 241 (2020) (prepublished online, November 2019).

Pape Møller, A. "Parallel Declines in Abundance of Insects and Insectivorous Birds in Denmark over 22 Years." *Ecology and Evolution* 9, no. 11 (2019).

Perry, G., et al. "Effects of Night Lights on Urban Reptiles and Amphibians." *Herpetological Conservation* 3 (2008).

Pettit, H., and Agence France-Presse. "Elephants Threatened by Poachers Are Evolving to Become Nocturnal So They Can Travel Safely at Night." *Daily Mail*, 13 September 2017.

Pinzon-Rodriguez, A., et al. "Expression Patterns of Cryptochrome Genes in Avian Retina Suggest Involvement of Cry4 in Light-Dependent Magnetoreception." *Journal of the Royal Society* 15, no. 140 (2018).

Pulgar, J., et al. "Endogenous Cycles, Activity Patterns and Energy Expenditure of an Intertidal Fish Is Modified by Artificial Light Pollution at Night (ALAN)." *Environmental Pollution* 244 (2019).

Raap, T., et al. "Light Pollution Disrupts Sleep in Free-Living Animals." *Scientific Reports* 5, no. 13557 (2015).

SOURCES

Rångtell, F. "If Only I Could Sleep—Maybe, I Could Remember." PhD diss., Institutionen för neurovetenskap, Uppsala universitet, 2019.

Restaurang Svartklubben. http://svartklubben.com.

Riccucci, M. "Lazzaro Spallanzani." *Bat Research News* 49, no. 4 (2008).

Riccucci, M., and J. Rydell. "Bats in the Florentine Renaissance: From Darkness to Enlightenment." *Lynx* 48 (2017).

Rich, C., and T. Longcore, eds. *Ecological Consequences of Artificial Night Lighting.* Washington, DC: Island Press, 2006.

Romanes, G. R. *Mental Evolution in Animals. With a Posthumous Essay on Instinct by Charles Darwin.* Cambridge: Cambridge University Press, 1883.

Russ, A., et al. "Seize the Night: European Blackbirds (*Turdus merula*) Extend Their Foraging Activity under Artificial Illumination." *Journal of Ornithology* 156 (2015).

Rybnikova, N., and B. A. Portnov. "Population-Level Study Links Short-Wavelength Nighttime Illumination with Breast Cancer Incidence in a Major Metropolitan Area." *Chronobiology International* 35, no. 9 (2018).

Rydell, J., et al. "Age of Enlightenment: Long-Term Effects of Outdoor Aesthetic Lights on Bats in Churches." *Royal Society Open Science* 4, no. 8 (2017).

Rydell, J., et al. "Dramatic Decline of Northern Bat *Eptesicus nilssonii* in Sweden over 30 Years." *Royal Society Open Science* 7, no. 2 (2020).

Salleh, A. "Light Pollution Delays Wallaby Reproduction and Puts Joeys at Risk." ABC News, 30 September 2015.

Sánchez-Bayo, F., and K. A. G. Wyckhuys. "Worldwide Decline of the Entomofauna: A Review of Its Drivers." *Biological Conservation* 232 (2019).

Sandberg, S. *Mørke—stjerner, redsel og fem netter på Finse.* Oslo, Norway: Samlaget, 2019.

Santos, C. D. "Effects of Artificial Illumination on the Nocturnal Foraging of Waders." *Acta Oecologica* 36, no. 2 (2010).

Saving Nemo Conservation Fund. www.savingnemo.org.

Scharping, N. "Why China's Artificial Moon Probably Won't Work." *Astronomy*, 26 October 2018.

Schoppert, S. "Emerging from the Darkness: 9 Creation Myths from Different Cultures." History Collection, 2017. https://historycollection.com/emerging-darkness-9-bizarre-creation-myths/.

Segedin, K. "Every Turtle Featured in Harrowing *Planet Earth II* Segment Was Saved." BBC Earth, 2016.

Sempler, K. "Den märkliga manicken från Antikythera." *Ny Teknik*, 3 March 2009.

———. "När Stockholm fick elektriskt ljus." *Ny Teknik*, 4 February 2018.

Singhal, R. K., et al. "Eco-physiological Responses of Artificial Night Light Pollution in Plants." *Russian Journal of Plant Physiology* 66 (2019).

Škvareninová, J., et al. "Effects of Light Pollution on Tree Phenology in the Urban Environment." *Moravian Geographical Reports* 25, no. 4 (2017).

SMHI Kunskapsbank. https://www.smhi.se/kunskapsbanken.

Smith, M. "Time to Turn Off the Lights." *Nature* 457, no 27 (2009).

Söderström, B. *Hur tänker din katt?* Stockholm: Bonnier Fakta, 2016.

Solly, M. "Swarms of Grasshoppers Invading Las Vegas Are Visible on Radar." *Smithsonian Magazine*, 30 July 2019.

"Some Like It Dark: Light Pollution and Salmon Survival." *Fish Report*, 4 June 2018.

Sperling, N. "The Disappearance of Darkness." In *Light Pollution, Radio Interference, and Space Debris, Astronomical Society of the Pacific Conference Series* 17 (1991), edited by L. Crawford.

Stagnelius, E. J. *Vän! I förödelsens stund.* Modernista (electronic edition), 2016.

Steinbach, R. "The Effect of Reduced Street Lighting on Road Casualties and Crime in England and Wales: Controlled Interrupted Time Series Analysis." *Journal of Epidemiology & Community Health* 69, no. 11 (2015).

Stevens, R. G. "What Rising Light Pollution Means for Our Health." *BBC Future*, 2016.

Stone, T. "Light Pollution: A Case Study in Framing an Environmental Problem." *Ethics, Policy & Environment* 20, no. 3 (2017).

———. "The Value of Darkness: A Moral Framework for Urban Nighttime Lighting." *Science Engineering Ethics* 24 (2018).

Strindberg, A. *Om det allmänna missnöjet, dess orsaker och botemedel.* Stockholm: Albert Bonniers Förlag, 1884.

Strömdahl, H. "Candela—grundenheten för grundstorheten ljusstyrka." *Kemivärlden Biotech med Kemisk Tidskrift* 2 (2015).

Svensson, A. M., and J. Rydell. "Mercury Vapour Lamps Interfere with the Bat Defence of Tympanate Moths (*Operophtera* spp.; Geometridae)." *Animal Behaviour* 55, no. 1 (1998).

SOURCES

Svensson, A. M., et al. "Avoidance of Bats by Water Striders (*Aquarius najas*, Hemiptera)." *Hydrobiologia* 489 (2002).

Svensson, P. *Ålevangeliet: Berättelsen om världens mest gåtfulla fisk.* Stockholm: Albert Bonniers Förlag, 2019.

Sveriges Radio P4 Sjuhärad. "Ljuset är nästa miljökatastrof." 9 January 2018.

Sveriges Radio P1. "Fysikpristagaren Peebles tog fram universums recept." *Vetandets värld*, 5 December 2019.

Sveriges Radio P1. "Så påverkas naturen av allt mer ljus." *Naturmorgon*, 30 December 2017.

Sveriges Radio P1. "Utflykter i natten—och älgar som far illa av varma somrar." *Naturmorgon*, 30 November 2019.

Sveriges Radio P1. "Vem äger mörkret?" *Vetenskapsradion Klotet*, 19 August 2015.

SVT Nyheter. "Mångalen?—Inte så galet." 11 November 2011.

SVT Nyheter. "Upplysta städer knäcker nattdjur." 8 July 2019.

Tähkämö, L., et al. "Systematic Review of Light Exposure Impact on Human Circadian Rhythm." *Chronobiology International* 36, no. 2 (2018).

Tanizaki, J. *Till skuggornas lov.* Reissue. Malmö, Sweden: ellerströms förlag, 1998.

Thomson, G. *Encyclopedia of Religion: Light and Darkness.* New York: Macmillan Reference, 2005.

Tidskriften Pilgrim 1 (2009).

Touzot, M., et al. "Artificial Light at Night Disturbs the Activity and Energy Allocation of the Common Toad during the Breeding Period." *Conservation Physiology* 7, no. 1 (2019).

Tracy Aviary. https://tracyaviary.org.

Tycho Brahe Museum. https://www.landskrona.se/en/se-gora/kultur-noje/museerochkonsthall/the-tycho-brahe-museum/.

Van Doren, B. M., et al. "High-Intensity Urban Light Installation Dramatically Alters Nocturnal Bird Migration." *Proceedings of the National Academy of Sciences* 114, no 42 (2017).

van Langervelde, F., et al. "Declines in Moth Populations Stress the Need for Conserving Dark Nights." *Global Change Biology* 24, no. 3 (2018).

Vogel, G. "Where Have All the Insects Gone?" *Science*, 10 May 2017.

SOURCES

Voigt, C. C., et al. "Guidelines for Consideration of Bats in Lighting Projects." EUROBATS Publication Series no. 8. Nairobi, Kenya: UNEP, 2018.

Wallace, D. R. *Beasts of Eden: Walking Whales, Dawn Horses, and Other Enigmas of Mammal Evolution.* Berkeley: University of California Press, 2004.

Westerdahl, C. "Kulturhistoria och grottor." *Svenska grottor* 5. Sveriges Speleologförbund, 1982.

Wheeling, K. "Artificial Light May Alter Underwater Ecosystems." *Science*, 28 April 2015.

Whyte, C. "Light Pollution's Effects on Birds May Help to Spread West Nile Virus." *New Scientist*, 24 July 2019.

Winger, B. M., et al. "Nocturnal Flight-Calling Behavior Predicts Vulnerability to Artificial Light in Migratory Birds." *Proceedings of the Royal Society B* 286 (2019).

Wistrand, S. "Eugène Jansson—inte bara blåmålare." *Kulturdelen*, 26 June 2015.

Witherington, B. E., et al. "Understanding, Assessing, and Resolving Light-Pollution Problems on Sea Turtle Nesting Beaches." *Technical Report TR2.* Tallahassee: Florida Fish and Wildlife Conservation Commission, 2014.

Wright Jr, K. P., et al. "Entrainment of the Human Circadian Clock to the Natural Light-Dark Cycle." *Current Biology* 23 (2013).

Zachos, E. "Too Much Light at Night Causes Spring to Come Early." *National Geographic*, 28 June 2016.

Zdanowicz, C. "Hordes of Grasshoppers Have Invaded Las Vegas." CNN, 27 July 2019.

Zhang, Z. "Man-Made Moon to Shed Light on Chengdu in 2020." *China Daily*, 19 October 2020.

Zubidat, A. E., and A. Haim. "Artificial Light at Night: A Novel Lifestyle Risk Factor for Metabolic Disorder and Cancer Morbidity." *Journal of Basic and Clinical Physiology and Pharmacology* 28, no. 4 (2017).

Index

INDEX

INDEX

About the Authors

Johan Eklöf, PhD, is a Swedish bat scientist and writer, best known for his work on microbat vision and, more recently, light pollution. He lives in the west of Sweden, where he works as a conservationist and copywriter. Johan has studied bats for almost twenty years and now has his own consultancy company, hired by authorities, wind companies, municipalities, city planners, and environmental organizations for expertise on bats, night ecology, and nature-friendly lighting. *The Darkness Manifesto* is his second book to be translated into English.

Elizabeth DeNoma is a freelance editor, translator, and publishing consultant. She completed her PhD in Scandinavian languages and literature at the University of Washington (Seattle) and taught Swedish language and composition for several years at the University of Wisconsin (Madison). She's the recipient of a Fulbright scholarship and the Swedish Women's Educational Award, sits on the advisory board of the University of Washington's Translation Studies Hub, and founded the National Nordic Museum's Meet the Author series, a monthly program showcasing Nordic authors in translation.